Discrete Choice Experiments Using R

Liang Shang · Yanto Chandra

Discrete Choice Experiments Using R

A How-To Guide for Social and Managerial Sciences

Springer

Liang Shang
Department of Social Sciences and Policy Studies
The Education University of Hong Kong
Tai Po, Hong Kong

Yanto Chandra
Department of Public and International Affairs
City University of Hong Kong
Hung Hom, Hong Kong

ISBN 978-981-99-4561-0 ISBN 978-981-99-4562-7 (eBook)
https://doi.org/10.1007/978-981-99-4562-7

This Springer imprint is published by the registered company Springer Nature Singapore Pte Ltd.
The registered company address is: 152 Beach Road, #21-01/04 Gateway East, Singapore 189721, Singapore

Preface

This book is a history and testament of our research journey as we evolve and grow as scholars. I (Yanto Chandra) was exposed to Discrete Choice Experiments (DCE) from many years of interactions with scholars from mathematical and operations sciences who have used DCE, econometrics, and R programming heavily in their research. After many courses in R and experiments, soul searching, and trial and error, I persuaded Shang to test drive the DCE method using R for her Ph.D. dissertation. We saw a lack of understanding and use of DCE in top and strong journals in our fields as an opportunity to further advance and disseminate this promising methodology. Building on her passion and talent in quantitative methods and R, Shang ended up producing one of the earliest Ph.D. dissertations that uses DCE in entrepreneurship research. Her research using DCE is expanding. Accordingly, this book is a journey of over 12 years or so in reading, discussing, experimenting, and implementing DCE research.

This book is also born out of our interest in studying how people make decisions and choices. Everyone makes a choice in their personal and professional life, from choosing what transport to use each day, which university to go to, which job to apply and accept, to which products to purchase, to which assets to invest in. Yet, mainstream scholars in the broader social sciences and business have not shown much interest in the study of choice behavior (except for some in marketing). This micro-level study of choices cannot be studied using conventional surveys or qualitative methods. It can be studied using laboratory or field experiments but research in this methodological tradition can be expensive and difficult to do and at times cannot capture the subtleties of choice behavior (e.g., choices that require multiple variables at different levels, where choices are stated than revealed) and where a choice is not independent of other alternatives. This is a welcoming opportunity to make a worthwhile contribution to unpack the mystery of DCE as a quasi-experimental methodology for researchers and analysts.

We design and structure the book in the most logical manner to help readers grasp the essence of DCE in a "hold-your-hands" approach. The book does not assume any prior knowledge in DCE or R programming but some knowledge of regression and coding will definitely be useful.

In Chap. 1, we open the conversation about DCE by discussing the distinction between stated preference methods vs revealed preference methods. We offered a quick brush of the history of these methods and offered a broad picture of DCE. We continued with Chap. 2, where we drilled down deeper into the subtleties of stated preference methods, including contingent valuation and choice modeling (in which DCE is a part of). In Chap. 3, we zoomed in on what DCE is, its theoretical foundation, key elements of DCE using practical examples, and the broad procedures in designing and implementing DCE, from identifying attributes to data analysis. We also offered a quick look at published research that has used DCE across different disciplines, from transportation, healthcare, and tourism to environment and key empirical aspects.

To orient readers with R that is used to design and implement DCE, we offered a description of R and its applicability for DCE in Chap. 4. Thus, this book assumes that readers do not have prior knowledge of the R programming environment. We offered step-by-step guidance on how to install R and RStudio, important packages in R to design DCE studies.

We further drilled down into DCE in Chap. 5 where we provided a sketch of the process in designing DCE research by focusing on asking the right type of research question. In Chap. 6, we offered a systematic guide on how to identify DCE attributes and levels from the literature, interviews, and expert opinions to social media analysis. We demonstrated this using an example of our demonstrative project on the alternative meat preference of consumers.

In Chap. 7, we demonstrated how to design DCE choice sets in R, particularly the full factorial and fractional factorial designs, the meaning of balanced and orthogonal designs, blocked fractional factorial designs, to the issue of labeled and unlabeled DCE designs. We showed how several key packages in R such as **support.CEs**, **idefix**, and **DCEtool** can be implemented to generate the attributes and levels for our demonstrative project.

In Chap. 8, we showed the techniques to design DCE choice sets in the questionnaire format using different R packages. We also provided illustrations on how to introduce a DCE study to respondents up to collecting their feedback prior to actual data collection.

In Chap. 9, we demonstrated how to use R to analyze DCE data, from data preparation and transformation, model estimation using various R packages (e.g., **survival** and **mlogit**) to calculating the willingness to pay values.

Finally, we demonstrated techniques to visualize the DCE data and report the results in Chap. 10, from generating data plots (e.g., box plots, histogram, correlations, Q-Q plot) to producing statistical outputs for publication purposes (e.g., descriptive statistics table, main effects analysis, and interaction analysis).

We concluded the book with a summary of what DCE is about while also suggesting some caveats to the DCE methodology. While DCEs offer several advantages over conventional research methods, they are not without limitations and potential pitfalls. As such, we have also highlighted some caveats to keep in mind when designing and implementing DCEs, such as the importance of sample size, choice of attributes and levels, and potential sources of bias.

We hope this book will become a valuable resource for anyone interested in stated preference methods, particularly discrete choice experiments, as an alternative to conventional research methods.

Tuen Mun, Hong Kong Liang Shang
Hung Hom, Hong Kong Yanto Chandra

Overview of Chapters

In Chap. 1, we broadly review stated preference methods and techniques. We introduce what stated preference methods are and their origins. We then compare and contrast revealed preference versus stated preference methods. In this chapter and throughout the book, we will focus on the choice-based approach (a type of stated preference method) that is also more popularly known as the *Discrete Choice Experiment* (DCE) approach.

In Chap. 2, we review the varieties of stated preference methods and discuss two major types, including contingent valuation and choice modeling. In particular, it highlights the differences between the discrete choice experimental (DCE) method and other choice modeling methods. In this chapter, we demonstrate how these stated preferences methodologies differ and why the differences and similarities matter to researchers interested in applying any of the methods. We also highlight the advantages of using choice modeling techniques for social and behavioral sciences research.

In Chap. 3, we discuss the fundamental elements, logic, and key principles in designing DCE studies. Accordingly, this chapter will be rather technical as we cover the nuts and bolts of the DCE methodology in detail.

In Chap. 4, we review the role of R as a computing platform and introduce various R packages to support the construction and implementation of DCE experiments. We discuss the benefits of using R and R packages as an open-source software to support DCE studies. We also provide brief examples of how to install and operate R for new R users.

In Chap. 5, we discuss how to define a specific research question and ground it in existing theories or literature. DCE experiments should ideally be grounded on a research question or puzzles. Puzzles can mean an unresolved research question or a question that is still debated by scholars We show how research questions can be developed using easy-to-follow examples to suit the DCE experiment.

In Chap. 6, we discuss the principles, approaches, and techniques to select an appropriate number of attributes and attribute levels using an actual research project that we did. We provide step-by-step instructions on how to establish attributes and levels in a DCE study.

In Chap. 7, we demonstrate techniques and steps to generate an orthogonal and balanced fractional factorial DCE design using R packages. The construction of experimental design is key to the successful implementation of DCE research. A full factorial design is the easiest design to create and use. It consists of all possible combinations of attribute levels but it is often considered too costly and tedious for respondents. A recommended approach is to employ fractional factorial design in the DCE design, which uses a randomized but smaller combination of attribute levels.

In Chap. 8, we discuss the techniques to design and construct a DCE questionnaire and collect data using survey software such as Qualtrics. Issues such as sampling frame, choices of sampling approach, minimum sample size, and questionnaire administration will also be discussed. This follows the best practices in generating a design matrix using various R packages that we discussed in Chap. 7.

In Chap. 9, we discuss and demonstrate the techniques to analyze the choice data generated, starting from data preparation, model selection to model specification and estimation. This demonstration involves the use of various R packages for estimating the classical multinominal logit model, calculating the goodness of fit, and computing the marginal willingness to pay values (i.e., relative importance of non-monetary attributes in DCE studies).

In Chap. 10, we show the techniques to visualize the results of DCE experiments using R and how to interpret the findings.

Chapter 11 offers a conclusion with our reflection on the advantages and shortfalls of the discrete choice experiments. We suggest that researchers can incorporate DCEs with other methods using a mixed method design to increase the credibility and validity of the findings.

Contents

About the Authors

Liang Shang is currently working as Assistant Professor in the Department of Social Sciences and Policy Studies at the Education University of Hong Kong. She obtained her Ph.D. in Public Policy from the City University of Hong Kong in 2020 and was awarded Sir Edward Youde Memorial Fellowship for Postgraduate Research Students in 2019. Prior to joining the Education University, she worked as Research Assistant Professor at Lingnan University and as Postdoctoral Fellow at the Chinese University of Hong Kong. With a primary research interest centered around the intersection of ethics and entrepreneurship, she focuses on examining the roles, practices, and mechanisms of business ventures with a strong overarching social purpose in contributing to social and environmental sustainability. Beyond this focus, she also has a keen interest in exploring emerging methods in social sciences, such as discrete choice experiments and artificial intelligence. Her work has been published in international journals, including Journal of Social Policy, Journal of Business Venturing Insights, and VOLUNTAS, among others.

Yanto Chandra Associate Professor of Management and Policy and Director of Technology, Policy, and Law Lab at the City University of Hong Kong. His research focuses on entrepreneurship and strategy in the technology and social contexts and their implications on policy, economy, and society. These include theory building, theory testing and technical explorations of artificial intelligence, blockchain and digital currencies. He is best-known for his work on social ventures, international new ventures, and more recently on decentralized autonomous organizations and the use of ChatGPT as data and method. Methods of interest include computer-aided text analysis, corpus linguistics, text mining, visual methods, image analysis, and data science using R. He is Associate Editor of *Journal of Business Venturing Insights* and several other journals in management. His work has been published in *Journal of International Business Studies, Journal of Business Venturing, Journal of Business Ethics, Journal of Business Venturing Insights, Journal of World Business,* as well as *Public Management Review, Public Administration Review,* and *World Development,* among others. His research has won international accolades including

Best Paper Award 2022 in the Academy of Management Annual Meeting and the Kooiman Prize for Best Paper in 2021.

Chapter 1
An Overview of Stated Preference Methods: What and Why

Stated preference methods are economic evaluation tools that allow scholars, practitioners, and policymakers to understand the preference and priority of decision-makers (e.g., consumers, citizens, investors, entrepreneurs, parents, or patients) in relation to the evaluation of a variety of goods or services or ideas or policies. Broadly speaking, stated preference methods involve collecting data on respondents' preferences and estimating *changes in utility* associated with a proposed change in quality or quantity of a product or service or idea or policy by *direct elicitation*. With direct elicitation, we mean that respondents are requested to evaluate multiple attributes or choose from various options (Adamowicz et al., 1994; Fujii & Gärling, 2003) thus, researchers can directly assess the utility of the choices.

A *choice set* (or *choice question*) is a situation where a decision-maker has to choose between two or more *alternatives* (e.g., choosing between buying car A or car B that vary between color, price, fuel efficiency, and country of origin attributes) or "Neither Car A or B". An option typically contains descriptions of hypothetical (e.g., new-to-the-world products or services) or existing (e.g., existing products or services) situations predefined by the researchers in a systematic and comprehensive manner. Due to their nature, stated preference methods are usually administered using *surveys* (i.e., via the web, phone, or mail; similar to conventional surveys except that it asks respondents to make "choices" among alternatives) to participants to capture how individuals or organizations would behave given changes in the products or services or ideas or policy of interest.

Compared with *revealed preference methods* that rely on market information to elicit consumer preference (e.g., based on the record of sales to identify factors important to customers when buying a car), stated preference methods capture and measure the *relative utility* (or importance or weight) of each attribute that forms various *alternatives* that describe a good (e.g., the relative importance of attributes of a car such as color vs price vs fuel efficiency vs country of origin).

L. Shang and Y. Chandra, *Discrete Choice Experiments Using R*,
https://doi.org/10.1007/978-981-99-4562-7_1

In this chapter, we will provide an overview of the stated preference methods by briefly reviewing the historical development of stated preference methods. Then, we will compare and contrast stated preference methods against revealed preference methods. We will also outline various forms of stated preference methods.

1.1 A Brief History of Stated Preference Methods

Stated preference methods refer to a family of techniques that can elicit an individual's preference toward a set of options via *surveys* (i.e., questionnaires containing close-ended questions) to estimate utility functions by using individual respondents' statements about their preference (Adamowicz et al., 1994). The basic conceptual framework for stated preference methods can be traced back to Thurstone's (1927, 1931) experimental and statistical work on measuring and scaling preferences. He conducted research to determine individuals' indifference curves and reported an experiment where participants were asked to make choices between different commodity bundles that consisted of (1) hats vs coats, (2) hats vs shoes, or (3) shoes vs coats (Thurstone, 1931). However, the responses in Thurstone's experiment were not sufficient to be used to predict market behavior (Ben-Akiva et al., 2019; Moscati, 2007). In the next 40 years following Thurstone's experiment, consumer experiments were largely limited to testing axioms for choice under uncertainty with no systematic attempt to integrate stated preference methods in demand analysis (Ben-Akiva et al., 2019; McFadden, 2017).

Stated preference methods grew in popularity in marketing research with commercial applications as early as the 1970s (Krantz & Tversky, 1971; Kroes & Sheldon, 1988). Since then, environmental economists have widely used the method to value the non-market benefit[1] of environmental change (see Green & Srinivasan, 1990). Luce and Tukey (1964) pioneered the first systematic use of stated preference methods, followed by the development of *conjoint analysis*[2] by market researchers such as Johnson (1974) and Louviere (1988), who applied stated preference methods to study consumer preference among familiar market products. A key feature of conjoint analysis is using experimental designs that allow at least a limited mapping of the preference of each respondent and the use of multiple measurements that allow elicitation of preference to be tested for consistency (Ben-Akiva et al., 2019). A good overview of the conjoint analysis method is provided by Green and Srinivasan (1978, p. 104), who also provided a formal definition for stated evaluation techniques: "*Any decompositional method that estimates the structure of a consumer's preference... given his/her overall evaluation of a set of alternatives that are pre-specified in terms*

[1] Many environmental goods, such as clean water and forests, are not traded in markets, and thus their value cannot be found in the marketplace.

[2] Conjoint analysis is a type of stated preference method that evaluates the product configurations independently of each other. We will discuss its differences with discrete choice experiments in Chap. 2.

of levels of different attributes". In addition, Cattin and Wittink (1982) conducted a survey to research firms to understand the commercial applications of conjoint analysis, including the response model for collecting preference judgments (i.e., rank order vs graded paired comparison vs rating scale vs others).

Around the 1980s, Louviere and Hensher introduced the use of stated preference experiments in their study that incorporated *choice-based* conjoint elicitations which directly mimicked market choice tasks (Louviere & Hensher, 1983). This choice-based experiment later evolved into what we have known as *discrete choice experiments (DCEs)*. In these experiments, respondents would be presented with a series of menus of products or services, and each would be described in terms of different attributes (e.g., price) and attribute levels. Respondents are then asked to choose their preferred product or service in each menu. The data produced from this kind of survey were easier to analyze and allowed greater prediction of the product or service's market shares.

1.2 Revealed Preference Versus Stated Preference

Studies of consumer (or citizen or patient or investor or entrepreneur; thus we use "consumer" in a very general sense) preference typically fall into two types: the revealed preference methods and stated preference methods. *Revealed preference methods* focus on studying past or present *actual* individual behaviors or choices to infer something about individuals' preference for a product, service, or a certain outcome (Boxall et al., 1996). Revealed preference data are collected either through *direct observations* on individual behavior (e.g., observing how consumers choose apples to buy in a shop) or using *surveys* to enquire about actual behavior (e.g., "Which factors are important for you when purchasing apples? Please tick or rank these factors: Sweetness, Brand, Color, Price, Country of Origin"). In contrast, in stated preference methods, researchers ask individuals questions (normally through surveys) that are intended to *elicit* their preference *without requiring them to act accordingly* (Kroes & Sheldon, 1988). We will show this in more vivid examples later.

Revealed preference methods are based on the revealed preference theory (Samuelson, 1938), which suggests that an individual's behavior could be seen as a series of choices. An individual's preference could be inferred by comparing *observed behavior* with available alternatives (e.g., choosing one brand of apple over other brands of apples). For instance, a study conducted by Avery et al. (2013) proposed a new ranking of U.S. undergraduate programs based on college students' revealed preferences. The authors argued that, when a student chooses a college among those that have offered him or her a place, that college "wins" and is ranked highest among other colleges. This study distributed a questionnaire to 3,240 high-achieving high school students who were likely to get accepted by many colleges. This study established preference-based rankings of colleges by capturing information on admission outcomes, scholarship offers, and students' final decisions (Avery et al., 2013). Therefore, the revealed preference approach is an appropriate and reliable tool for deriving utilities and estimating models of consumer demand as it records what people *actually do*, rather than what they say they will do (Ben-Akiva et al., 1994; Wardman, 1988).

However, revealed preference methods have several drawbacks, making them impractical. First, revealed preference methods rely on *variation* in the real-world alternatives so that the statistical model can investigate how various factors important for describing the product or service of interest could influence individuals' choices. For instance, if most students from the study by Avery et al. (2013) only received one or two college offers, it would be impossible to model the different effects of the factors (e.g., offers by other colleges with their attributes such as tuition fees, location, ranking, etc.). Thus, they cannot capture *counterfactuals* (e.g., what choices would have been made by an individual when other alternatives were present).

Secondly, there are often *strong correlations* between explanatory variables of interest (for example, side effects of medicine and compliance rates) (Getz & Huang, 1978; Mark & Swait, 2004), making it difficult to elicit which factor is influencing individuals' choices and thus affecting the accuracy of revealed preference methods. Thus, revealed preference studies suffer from *multicollinearity issues* between explanatory variables (Earnhart, 2001; Kroes & Sheldon, 1988). However, isolation is sometimes required and important if researchers are only interested in valuing an improvement in one of the attributes of a product or service (Adamowicz et al., 1994).

Thirdly, given that revealed preference methods capture and measure choices based on *actual behavior*, it will be difficult to *forecast* demand for new services under conditions that do not yet exist (Ben-Akiva et al., 1994; Kroes & Sheldon, 1988; Wardman, 1988). In fact, forecasting and predicting are paramount for better decision-making. For example, it is not possible to forecast market demand for an apartment on planet Mars as it does not exist, and there are no similar products like this on the market.

On the other hand, *stated preference methods* can overcome the challenges above and have become an attractive option in social and behavioral science research. In particular, stated preference methods can help researchers identify behavioral responses of individuals to *hypothetical or new* (as well as *existing*) situations which have not yet been revealed on the market (Boxall et al., 1996). Individuals' preferences for *existing* products or services can also be studied, such as the study of choices of public health facilities by patients that involve existing modes and current practices of healthcare services (Honda et al., 2015).

Compared with revealed preference methods, the stated preference methods can produce more accurate results because researchers *predefine* the conditions (e.g., attributes and their levels) being evaluated by respondents (Kroes & Sheldon, 1988) while also giving individuals the freedom not to choose from the alternatives presented to them (e.g., the "Neither A or B" option). Stated preference methods also offer more *flexibility* as researchers can include *a wide variety of variables (comprising a mix of new or hypothetical and existing attributes)* of interest in designing the experiments. Moreover, stated preference methods are effective means to *measure non-use values*, which are values assigned by individuals to a product or service that do not concern current or future uses (Champ et al., 1997; McFadden, 2017).

Nonetheless, stated preference methods also have some drawbacks. The first is that individuals are likely to *overstate their valuation* of a particular product, service, or outcome, leading to *biased estimates* of relative value (or "hypothetical bias") (Johansson-Stenman & Svedsäter, 2012). But this (social desirability) bias can also occur in surveys, lab experiments, or any revealed preference studies because people's behavior can change when prompted or measured. Furthermore, stated preference methods have often been criticized for their hypothetical nature in the research design and the fact that the actual behavior of respondents is not observed. Because respondents may not necessarily do what they indicate in surveys (Kroes & Sheldon, 1988), stated preference methods require carefully designed surveys that are grounded in real-world information and sophisticated data analysis, hence they could be more *time- and resource-consuming* (e.g., Ben-Akiva et al., 1994) and demand *higher computing skills.* However, the advantages of stated preference methods outweigh their potential drawbacks in terms of untangling the preference structure of each individual and the overall individuals in a study in a relatively *efficient* manner using *scientifically rigorous* approaches.

These limitations can be addressed by using *mixed methods* that combine DCE and other methods. For example, one can use in-depth interviews to elicit attributes for use in a stated preference study or conduct qualitative research as a follow-up study to probe the results. Alternatively, researchers can address the limitations using *replication studies.* For instance, using two or more stated preference studies where study 1 sets the preliminary relationships among variables and study 2 confirms and extends it. There are many ways and new techniques that continue to evolve, further enhancing the capabilities of stated preference methods. After all, the same knife can create an artistic carving or not—it all depends on the craftsmanship and imagination of the artists.

Combining data on actual and intended behavior can also overcome the limitations of using a singular data type. For example, we are interested in managers' preference to use human-like A.I. robots. Researchers could combine revealed preference data on actual purchases of traditional robots and include new features of human-like A.I. robots (e.g., facial recognition, human appearance, natural conversation) to better estimate preference and potential success for such new products.

References

Adamowicz, W., Louviere, J., & Williams, M. (1994). Combining revealed and stated preference methods for valuing environmental amenities. *Journal of Environmental Economics and Management, 26*(3), 271–292.

Avery, C. N., Glickman, M. E., Hoxby, C. M., & Metrick, A. (2013). A revealed preference ranking of U.S. colleges and universities. *The Quarterly Journal of Economics, 128*(1), 425–467.

Ben-Akiva, M., Bradley, M., Morikawa, T., Benjamin, J., Novak, T., Oppewal, H., & Rao, V. (1994). Combining revealed and stated preference data. *Marketing Letters, 5*(4), 335–349.

Ben-Akiva, M., McFadden, D., & Train, K. (2019). Foundations of stated preference elicitation: Consumer behavior and choice-based conjoint analysis. *Foundations and Trends in Econometrics, 10*, 1–144.

Boxall, P. C., Adamowicz, W. L., Swait, J., Williams, M., & Louviere, J. (1996). A comparison of stated preference methods for environmental valuation. *Ecological Economics, 18*(3), 243–253.

Cattin, P., & Wittink, D. R. (1982). Commercial use of conjoint analysis: A survey. *Journal of Marketing, 46*(3), 44–53.

Champ, P. A., Bishop, R. C., Brown, T. C., & McCollum, D. W. (1997). Using donation mechanisms to value nonuse benefits from public goods. *Journal of Environmental Economics and Management, 33*(2), 151–162.

Earnhart, D. (2001). Combining revealed and stated preference methods to value environmental amenities at residential locations. *Land Economics, 77*(1), 12–29.

Fujii, S., & Gärling, T. (2003). Application of attitude theory for improved predictive accuracy of stated preference methods in travel demand analysis. *Transportation Research Part A: Policy and Practice, 37*(4), 389–402.

Getz, M., & Huang, Y. C. (1978). Consumer revealed preference for environmental goods. *The Review of Economics and Statistics*, 449–458.

Green, P. E., & Srinivasan, V. (1978). Conjoint analysis in consumer research: Issues and outlook. *Journal of Consumer Research, 5*(2), 103–123.

Green, P. E., & Srinivasan, V. (1990). Conjoint analysis in marketing: New developments with implications for research and practice. *Journal of Marketing, 54*(4), 3–19.

Honda, A., Ryan, M., van Niekerk, R., & McIntyre, D. (2015). Improving the public health sector in South Africa: Eliciting public preferences using a discrete choice experiment. *Health Policy and Planning, 30*(5), 600–611.

Johansson-Stenman, O., & Svedsäter, H. (2012). Self-image and valuation of moral goods: Stated versus actual willingness to pay. *Journal of Economic Behavior & Organization, 84*(3), 879–891.

Johnson, R. M. (1974). Trade-off analysis of consumer values. *Journal of Marketing Research, 11*(2), 121–127.

Krantz, D. H., & Tversky, A. (1971). Conjoint-measurement analysis of composition rules in psychology. *Psychological Review, 78*(2), 151–169.

Kroes, E. P., & Sheldon, R. J. (1988). Stated preference methods. *Journal of Transport Economics and Policy, 22*(1), 11–25.

Louviere, J. J. (1988). Conjoint analysis modelling of stated preference: A review of theory, methods, recent developments and external validity. *Journal of Transport Economics and Policy, 22*(1), 93–119.

Louviere, J. J., & Hensher, D. A. (1983). Using discrete choice models with experimental design data to forecast consumer demand for a unique cultural event. *Journal of Consumer Research, 10*(3), 348–361.

Luce, R. D., & Tukey, J. W. (1964). Simultaneous conjoint measurement: A new type of fundamental measurement. *Journal of Mathematical Psychology, 1*(1), 1–27.

Mark, T. L., & Swait, J. (2004). Using stated preference and revealed preference modeling to evaluate prescribing decisions. *Health Economics, 13*(6), 563–573.

McFadden, D. (2017). *Stated preference methods and their applicability to environmental use and non-use valuations*. Edward Elgar Publishing.

Moscati, I. (2007). Early experiments in consumer demand theory: 1930–1970. *History of Political Economy, 39*(3), 359.
Samuelson, P. A. (1938). A note on the pure theory of consumer's behaviour. *Economica, 5*(17), 61–71.
Thurstone, L. L. (1927). A law of comparative judgment. *Psychological Review, 34*(4), 273.
Thurstone, L. L. (1931). The indifference function. *The Journal of Social Psychology, 2*(2), 139–167.
Wardman, M. (1988). A comparison of revealed preference and stated preference models of travel behaviour. *Journal of Transport Economics and Policy*, 71–91.

Chapter 2
A Comparison of Stated Preference Methods

This chapter reviews the varieties of stated preference methods and discusses two major types, *contingent valuation* and *choice modeling*. In particular, it highlights the differences between the *discrete choice experimental* (*DCE*) method and other choice modeling methods. In this chapter, we demonstrate how these stated preferences methodologies differ and why the differences and similarities matter to researchers interested in applying any of the methods. We also highlight the advantages of using choice modeling techniques for social and managerial sciences research.

2.1 Stated Preference Methods and Total Economic Value

Stated preference methods refer to a family of techniques that use individual respondents' statements to *elicit their preferences* in a set of situations or contexts to estimate utility functions (i.e., importance) of products, services, ideas, or policies. Stated preference methods and revealed preference methods are both valuation techniques that are based on the economics of benefit measurement (Adamowicz et al., 1994). Stated preference methods are effective means to estimate both *use* and *non-use values*, which distinguish them from revealed preference methods that can only capture use values (Champ et al., 1997; Eom & Larson, 2006; McFadden, 2017). To better understand how stated preference methods work and what they do, we need to first understand what *economic value* is about.

Economic value refers to the value individuals place on goods or services based on their preferences and willingness to pay for them. In particular, the total economic value comprises the sum of use and non-use values. Use values may be *direct* (e.g., by using the service) or *indirect* (e.g., the benefit an individual might obtain from a good without actually using it directly). For example, individuals can derive direct use value from visiting a forest or eating vegetables (i.e., direct use value). They may also derive benefits from the existence of a beautiful forest near their house because it increases property value even though they never visit it (i.e., indirect use value).

L. Shang and Y. Chandra, *Discrete Choice Experiments Using R*,
https://doi.org/10.1007/978-981-99-4562-7_2

Additionally, individuals may have an *option value* for a product or service and be willing to pay to ensure or preserve access to future benefits of such products or services. If the option of future use relates to individuals' own use, such willingness to pay reflects the option value.

On the other hand, *non-use values*, also known as *passive values*, emerge in contexts where an individual is willing to pay for a product or service even though they make no direct use of it or may not derive benefits from it directly. Generally speaking, non-use values refer to those assigned by individuals to a resource that does not concern their current or future uses. Such values are determined by people and are measured by *people's willingness to pay to preserve the resource* (Adamowicz et al., 1994; Whitehead et al., 2008). These values might be derived from simply knowing a certain resource exists (i.e., *existence value*), even if it is or will never be utilized or consumed (Oglethorpe & Miliadou, 2000). For example, an individual may be willing to pay to preserve a natural resource like a waterfall for its existence rather than current or future use. Moreover, if individuals place value on passing on the resource or ensuring the availability of use to future generations rather than considering the current or future use for themselves, this is called a *bequest value* (Walsh et al., 1984). This value is driven by a desire to leave a legacy or ensure that future generations can experience the resource. We summarized the key characteristics of both use values and non-use values with examples in Table 2.1.

In general, non-use values are concerned with people's *willingness to pay* (WTP) for something that they do not consume themselves (Adamowicz et al., 1998). This concept has been widely discussed and applied in environmental economics (e.g., Huang et al., 1997) as individuals often hold positive value for many environmental

Table 2.1 A classification of economic value and its descriptions

Total economic value			
Classification of economic values		Description	Example
Use value	Direct use value	Values derived from directly using a product	Benefits gained from directly consuming food and vegetables
	Indirect use value	Benefits gained from other people's use of a product or being in close proximity to the product	Benefits gained from living close to a wetland because it provides important ecosystem services
	Option value	Values of keeping open the option to use a resource in the future	Values of preserving the public park for one's own future use
Non-use value	Existence value	Values that individuals attach to a product for simply knowing its continued existence	Willingness to pay for preserving an endangered butterfly species
	Bequest value	Values of preserving a resource for their future generations to enjoy	Willingness to pay for preserving a river to ensure one's children can swim in it in the future

products even without directly or indirectly using them (Wilson & Hoehn, 2006). Such values *cannot be directly observed* using revealed preference methods, which determine the values individuals place on a product or service by observing their purchase or consumption behaviors. For example, someone might value a forest because they want to preserve it for future generations, even if they never visit there. Since this person never actually uses the resource, their preference for preserving it cannot be revealed through their *observed behavior*. Therefore, stated preference methods have become feasible to *include passive use* considerations in the valuation analysis by using individuals' statements about their preferences to estimate the non-market and non-use benefits.

Due to their nature (i.e., asking individuals questions to elicit their preference), stated preference methods require purposefully and carefully designed *surveys* for data collection (Kroes & Sheldon, 1988).

Specifically, stated preference methods could be broadly classified into two major types: ***contingent valuation*** and ***choice modeling***. Contingent valuation (CV) is a technique to estimate the values that individuals place on a product or service by asking questions that help reveal the monetary trade-off each individual would make concerning the value of the product or service. Choice modeling (CM) differs from the CV approach in that CM asks individuals to *rank, rate,* or *select from* different alternatives from which the utility of a product or service can be inferred. The discrete choice experiment (DCE) we will focus on throughout this book belongs to the CM approach. The following sections will introduce each method in detail.

2.2 Contingent Valuation

2.2.1 Definition and Types of Contingent Valuation

Contingent valuation (CV) is a survey-based method to estimate the monetary valuation of *non-market* products, which are products and services not traded in markets and do not have a market price (e.g., clean air and water), by obtaining an individual's *willingness to pay* (WTP) for acquiring or improving the quality of a product or service; that is their *willingness to accept* (WTA) for a loss or degradation of the product or service (Hanemann, 1991; Shogren et al., 1994). This method is called *contingent* because its valuation of non-market products depends on *hypothetical scenarios* presented to individuals. A key advantage of CV is that the answers to WTP or WTA questions focus directly on the theoretically correct monetary measures of utility changes.

In particular, WTP is considered an appropriate measure of the *economic value* of a product or service when an individual wants to *obtain a certain product or service.* At the same time, WTA is more appropriate when an individual is *asked to give up a product or service voluntarily*. In this context, it is important to draw out either the maximum WTP or the minimum WTA to align with the underlying theory of economic valuation. CV attempts to elicit the value an individual attaches to a non-market good, such as environmental quality or cultural heritage, by asking a simple question in either form below:

- *What is the maximum you would be willing to pay to enjoy the benefits provided by the non-market good? or*
- *What is the minimum compensation you would be willing to accept to endure the loss or degradation of the good?*

A variety of question formats are available within CV to get individuals to indicate their WTP or WTA (Venkatachalam, 2004), where the four main forms include:

- Open-ended
- Bidding game elicitation
- Referendum
- Payment card.

Under the *open-ended* CV, a respondent is directly asked about the maximum amount an individual is willing to pay for benefits from the non-market good or the minimum amount *they are willing to accept to forgo the benefits from it*. An example is provided below.

An example of open-ended CV:
"What is the maximum amount that you would be prepared to pay every year to conserve public parks in your neighborhood?"
Answer:
____

In the *bidding game elicitation*, a respondent is asked if they are *willing to pay* a certain amount of money *in exchange for a positive change* (e.g., improving the landscape of a public park) in a product or service. If the respondent answers "yes", the interviewer will keep increasing the amount until a "no" answer is obtained. An example is provided below.

An example of a bidding game elicitation in CV:
"Are you willing to pay $20 per year for improving the landscape of the XYZ park?"
Yes ☐ No ☐
If respondents answer yes, then ask the following (repeating this until we get a No answer)
"Are you willing to pay $25 per year for improving the landscape of the XYZ park?"
Yes ☐ No ☐

Furthermore, under the *referendum* CV (also known as *dichotomous choice* or *take-it-or-leave-it* question), a respondent is randomly presented with differing $$$ amounts and then asked to vote if they are willing to pay that specific amount for a positive change in a product or service. This format has been seen as *mimicking what happens in a real market situation* in which a consumer will decide whether to buy or not after a seller gives a product or service a quotation. In this format, each respondent is asked only one question, whether they are prepared to pay a certain $$$ to be able to enjoy the benefits of the non-market commodity, as shown in the example below:

> *An example of a referendum CV*:
> "Would you be willing to pay $15 per year for urban green space conservation in the city?" (Take it or leave it)
> Yes ☐ No ☐

Lastly, a *payment card* CV presents a respondent with a card or a table with an array of payment values and asks them to select the value closest to the maximum amount they are willing to pay. The amount selected by respondents is then interpreted as their WTP. For instance:

> *An example of a payment card CV*:
> "What is the maximum amount you would pay per year for the conservation of wild animals? Please circle the dollar amount from the list below"
>
> | | | | | |
> |---|---|---|---|---|
> | $0 | $2 | $5 | $15 | $20 |
> | $30 | $40 | $50 | $80 | $100 |
>
> More than $100

2.2.2 Limitations of Contingent Valuation

Contingent valuation (CV) has been widely used as a valuation method to elicit both use and non-use values with its direct estimation of WTP or WTA to provide useful information about the values of a specific non-market good. Despite its wide application, this method has several *limitations*—first, the differences between WTP and WTA. In an ideal situation, if the WTP is only a small fraction of an individual's income (which means that paying the amount of money will not have too much influence on the person's life), there should be no income effect. Thus, WTP and WTA should be more or less equal. Nonetheless, many studies found that individuals' WTA is often larger than WTP for the same product or service (e.g., Bishop & Heberlein, 1979; Hammack & Brown, 2016). This divergence might be because people generally *value losses more heavily* than they value a gained benefit of the same magnitude (Hanemann, 1991). There might be an instant endowment effect (Kahneman et al.,

1990) in which individuals attach additional value to their possession of a certain product or service.

Secondly, respondents may sometimes have difficulty responding to a CV due to this method's nature of asking for individuals' WTP or WTA. Theoretically, a CV can be used to value different characteristics of a product or service as well as the whole product or service. However, in practice, if the research focus is on the economic value of different characteristics of the product or service in question, the CV questionnaire can become too burdensome and difficult to manage for respondents because they will need to evaluate a series of options and indicate their WTP or WTA one by one. Moreover, the referendum format of CV may be subject to a "yea-saying" response (Blamey et al., 1999), in which respondents tend to select "yes" to answer questions regardless of their content.

The third limitation associated with CV concerns respondents "protest votes". Many studies have found that a significant number of respondents to CV surveys tend to either *state a zero value or refuse to place a monetary value* on a non-market commodity at all (Jorgensen et al., 1999). Clearly, some people have trouble attaching direct monetary values to non-market products (e.g., clean air or safe neighborhood) for reasons associated with the process of CV itself (e.g., perceived unfairness of having to pay extra for a non-market commodity). These responses are viewed as threats to the validity of valuation. For instance, a person may regard an environmental good (e.g., clean air) as having values but still place a null value in CV surveys as a "protest" action; that is, they refuse the idea of placing monetary values on public products. Such protest votes can hardly be completely removed from the dataset and could lead to biased estimates in CV.

On the other hand, choice modeling (CM) may avoid some of the problems associated with CV since it asks for *ratings, rankings, or direct selection between two or more choices* instead of requiring respondents to think in monetary terms. The following section will introduce different formats of CM in detail.

2.3 Choice Modeling

While the contingent valuation (CV) method seeks to elicit values that an individual attaches to a product or service by directly asking questions about WTP or WTA, choice modeling (CM), which includes DCEs, seeks to capture and examine the rankings and ratings of *alternatives* inferred from individuals' preferences of a product or service. CM is a family of survey-based techniques for eliciting individuals' preferences for products, where products are described using their attributes and attribute levels (Hanley et al., 2001). CM method is based on the idea that a product or service or idea or policy can be characterized in terms of its attributes (i.e., salary, work culture, working hours, and location as attributes of a job offer) and the different levels of these attributes (e.g., friendly vs toxic as two levels of work culture; 8 vs 10 vs 12 h as three levels of working hours). This approach assumes that changing attribute levels will lead to generating a different product or service

(Rolfe et al., 2000). Similar to CV, CM could also measure both the use and non-use values of products or services. But instead of directly asking respondents to think of the monetary values of products, CM asks respondents to offer rankings, ratings, or choices between different alternatives. Thus, it could avoid some of the problems like protest votes associated with CV approaches. Generally, there are four major forms of CM: *contingent ranking, contingent rating, graded paired comparisons, and discrete choice experiments (DCEs).*

In the *contingent ranking* approach, a respondent is asked to *compare* and *rank* a series of alternatives from the most preferred to the least preferred, in which a number of attributes with varying attribute levels characterize the alternatives. Such an approach would yield the same information as would come from directly asking respondents to indicate the option that they feel is the best and the option that they feel is the worst (which is also known as the best–worst scaling approach) (Louviere et al., 2015). One limitation of this ranking approach is that it could increase the cognitive burden of respondents when there are a lot of alternatives for respondents to consider and rank based on their preferences. An example of contingent ranking is provided below:

An example of a contingent ranking:
"Rank the alternatives below to your preferences, by assigning 1 to the most preferred option and 3 to the least preferred"

	Coffee Beans A	Coffee Beans B	Coffee Beans C
Fair-trade products	Yes	No	Yes
Organic certification	No	Yes	Yes
Eco-friendly label	Yes	No	No
Price	$7/pound	$7/pound	$10/pound
Ranking (1, 2, 3)			

In the *contingent rating* approach, a respondent is also presented with a number of alternatives to choose from, but they are asked to rate these alternatives on a *semantic or numeric scale* instead of ranking them. This approach does not require respondents to compare alternatives directly but asks them to *evaluate each alternative respectively*. One of the major drawbacks of this approach is that different respondents may have different standards for rating an alternative. A rating of "6" given by a respondent might have a different meaning from the same "6" given by another respondent. An example of contingent rating is provided below:

An example of a contingent rating:
"On the scale below, please rate your preference for choosing the following coffee beans"

	Coffee Beans A
Fair-trade products	Yes
Organic certification	No
Eco-friendly label	Yes
Price	$7/pound
Rating: 1 2 3 4 5 6 7	8 9 10
Very low preference	Very high preference

A discrete choice experiment (DCE) could help address the abovementioned limitations with contingent ranking and rating. In a DCE, a respondent is presented with a set of alternatives, characterized by different attributes at varying levels, and is asked to select the alternative that is more or most preferred by them out of two or more alternatives at one go.

In particular, a DCE consists of three key steps. First, collecting data by asking respondents to consider *trade-offs* (i.e., prioritizing one over the other) among desirable alternatives. Second, a computational method to derive the *utilities* of the attributes (e.g., for color, price, fuel efficiency, etc.) with a high degree of accuracy for each respondent's choice behavior (e.g., Respondent #1 will likely place different utility for color, price, fuel efficiency compared to Respondent #52). Third, a simple market *simulation model* seeks to determine the attributes of a product or a service that will maximize its share of preference (Johnson, 1974). This approach draws its behavioral theory from a subset of decision theory that highlights the modeling of individual preference behaviors based on how individuals integrate information in the evaluation of *predefined alternatives* constructed by researchers (Anderson, 1981).

Louviere and Hensher (1982) originally developed the DCE approach and have since gained wide acceptance in market research (Blamey et al., 1999). An example of a choice experiment is shown below:

An example of a DCE:
"Please select which alternative you prefer".

	Coffee Beans A	Coffee Beans B
Fair-trade products	Yes	No
Organic certification	No	Yes
Eco-friendly label	Yes	No
Price	$7/pound	$7/pound

Which alternative do you prefer?
Coffee Beans A □
Coffee Beans B □
Choose Neither □

This approach uses a similar theoretical foundation as the referendum CV technique (in which respondents are asked to indicate whether they will take it or leave it after evaluating the option), as they are both based on *random utility theory* (Hanemann, 1984; Hanley et al., 1998; McFadden, 1974). We will discuss random utility theory in more detail in later chapters.

Furthermore, *graded paired comparisons-based* CM combines elements of DCEs in selecting the more or most preferred alternative and contingent rating by rating the strength of preference to each alternative. In graded paired comparisons CM, a respondent is presented with a set of two choices and then asked to select their preferred alternative and indicate the strength of their preference on a numeric or semantic scale. This approach seeks more information from respondents and enables relative estimates of the value of unit changes in the attributes. However, despite the paired comparison approach allowing researchers to capture more information from respondents, it may also increase the cognitive burden on respondents if there are many choice sets for them to evaluate.

An example of a graded paired comparison CM approach is presented below.

An example of a graded paired comparison:
"Which coffee beans do you prefer, given the two alternatives provided below? Please indicate your preference and the strength of your preference"

	Coffee Beans A	Coffee Beans B
Fair-trade products	Yes	No
Organic certification	No	Yes
Eco-friendly label	Yes	No
Price	$7/pound	$7/pound

1 2 3 4 5 6 7 8 9 10

Strongly prefer A Strongly prefer B

There have been long-standing debates among researchers about the advantages and disadvantages of contingent rating and ranking and DCEs (e.g., Boxall et al., 1996; Fischer & Hawkins, 1993; Moore, 2004). A fundamental difference between these methods is that the contingent rating and ranking use *judgments* to elicit preference, while a DCE uses *choices* (Karniouchina et al., 2009). In most cases, it is reasonable to assume that if a person rated or ranked one alternative higher than another on a structured ordinal scale, that person would also select the first alternative as the more preferred one over the second in a choice task. However, existing evidence has shown both descriptive and procedural variance between the two approaches (e.g., Bettman et al., 1998; Payne et al., 1992) as different response modes may trigger different response strategies, leading to different results in stated preference studies (Karniouchina et al., 2009). For example, prior research has indicated that respondents may implicitly consider a judgment task (i.e., rating or ranking) as a request to indicate their *personal satisfaction* while interpreting a choice task as a request to judge which of the options within each choice set is more *acceptable* to them (Bazerman et al., 1992). Form matters!

In general, we argue that *choice-based* approaches, including *DCEs,* resemble the *marketplace behavior* more than the profile rankings and ratings approaches. This is because choice tasks are *more realistic* and *mimic* the decisions that consumers perform on an everyday basis.[1] Thus, scholars argue that choice-based approaches provide *greater external validity* (Elrod et al., 1992). Moreover, choice-based approaches constitute *easier tasks* for respondents to answer because they can make each choice independently (within a limited set of alternatives of two to three in each choice set) without worrying about ranking or rating scales consistently over different alternatives. In particular, in ranking or rating tasks, respondents must worry whether their indicated preference is based on the same standards for all alternatives. Hence this can be cognitively demanding when many alternatives are being evaluated.

In summary, CM is a family of survey-based techniques that measure a product or service's use and non-use values by eliciting respondents' preferences. Both CM and CV belong to stated preference methods but with some differences (the family tree of the stated preference method is summarized in Fig. 2.1), as discussed in this chapter.

CM differs from the CV approach as CM avoids directly asking respondents to consider the value of a product or service in monetary terms. Thus, the CM technique can potentially address some challenges and limitations of CM. In particular, CM allows estimating the value of each attribute that makes up a product or service and situational changes, given that the CM technique is based on attributes (Hanley et al., 1998). Moreover, CM could avoid the "yea-saying" problem of the referendum CV since respondents would not face "all or nothing" choices.

---

[1] Buying a mobile phone is a discrete 0 or 1 choice by selecting or evaluating a group of competing mobile phones that vary across brand, country of origin, design, color, price, etc.; rather than ranking all possible options to make a purchase decision.

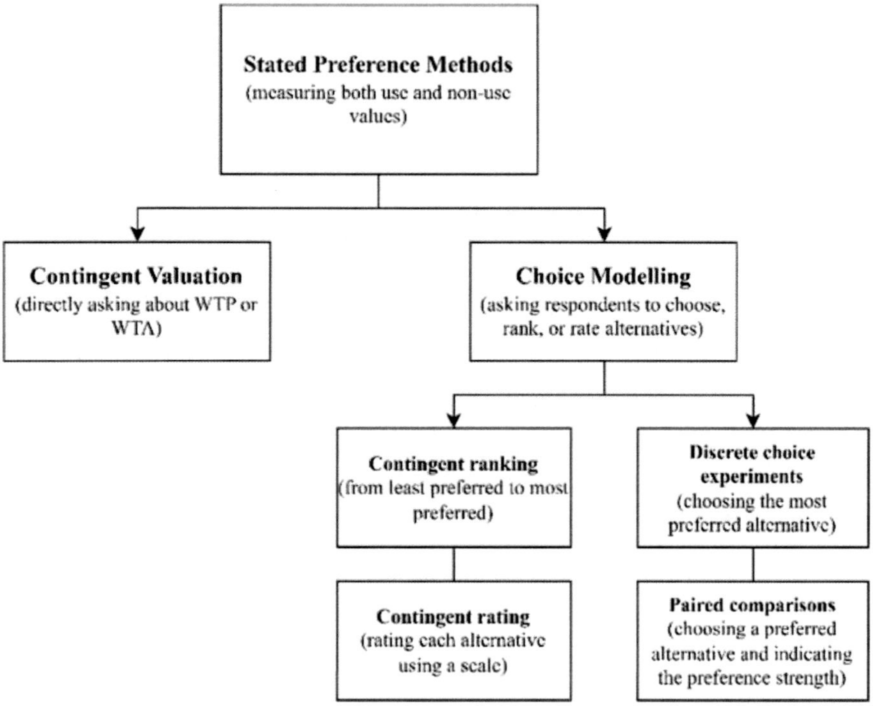

Fig. 2.1 A summary of stated preference methds

References

Adamowicz, W., Louviere, J., & Williams, M. (1994). Combining revealed and stated preference methods for valuing environmental amenities. *Journal of Environmental Economics and Management, 26*(3), 271–292.

Adamowicz, W., Boxall, P., Williams, M., & Louviere, J. (1998). Stated preference approaches for measuring passive use values: Choice experiments and contingent valuation. *American Journal of Agricultural Economics, 80*(1), 64–75.

Anderson, N. H. (1981). *Foundations of information integration theory*. Academic Press.

Bazerman, M. H., Loewenstein, G. F., & White, S. B. (1992). Reversals of preference in allocation decisions: Judging an alternative versus choosing among alternatives. *Administrative science quarterly*, 220–240.

Bettman, J. R., Luce, M. F., & Payne, J. W. (1998). Constructive consumer choice processes. *Journal of Consumer Research, 25*(3), 187–217.

Bishop, R. C., & Heberlein, T. A. (1979). Measuring values of extramarket products: Are indirect measures biased? *American Journal of Agricultural Economics, 61*(5), 926–930.

Blamey, R. K., Bennett, J. W., & Morrison, M. D. (1999). Yea-saying in contingent valuation surveys. *Land Economics, 75*(1), 126–127.

Boxall, P. C., Adamowicz, W. L., Swait, J., Williams, M., & Louviere, J. (1996). A comparison of stated preference methods for environmental valuation. *Ecological Economics, 18*(3), 243–253.

Champ, P. A., Bishop, R. C., Brown, T. C., & McCollum, D. W. (1997). Using donation mechanisms to value non-use benefits from public products. *Journal of Environmental Economics and Management, 33*(2), 151–162.

Elrod, T., Louviere, J. J., & Davey, K. S. (1992). An empirical comparison of ratings-based and choice-based conjoint models. *Journal of Marketing Research, 29*(3), 368–377.

Eom, Y. S., & Larson, D. M. (2006). Improving environmental valuation estimates through consistent use of revealed and stated preference information. *Journal of Environmental Economics and Management, 52*(1), 501–516.

Fischer, G. W., & Hawkins, S. A. (1993). Strategy compatibility, scale compatibility, and the prominence effect. *Journal of Experimental Psychology: Human Perception and Performance, 19*(3), 580.

Hammack, J., & Brown Jr, G. M. (2016). *Waterfowl and wetlands: Toward bioeconomic analysis*. Routledge.

Hanemann, W. M. (1984). Welfare evaluations in contingent valuation experiments with discrete responses. *American Journal of Agricultural Economics, 66*(3), 332–341.

Hanemann, W. M. (1991). Willingness to pay and willingness to accept: How much can they differ? *The American Economic Review, 81*(3), 635–647.

Hanley, N., Mourato, S., & Wright, R. E. (2001). Choice modelling approaches: A superior alternative for environmental valuation? *Journal of Economic Surveys, 15*(3), 435–462.

Hanley, N., Wright, R. E., & Adamowicz, V. (1998). Using choice experiments to value the environment. *Environmental and Resource Economics, 11*(3), 413–428.

Huang, J. C., Haab, T. C., & Whitehead, J. C. (1997). Willingness to pay for quality improvements: Should revealed and stated preference data be combined? *Journal of Environmental Economics and Management, 34*(3), 240–255.

Johnson, R. M. (1974). Trade-off analysis of consumer values. *Journal of Marketing Research, 11*(2), 121–127.

Jorgensen, B. S., Syme, G. J., Bishop, B. J., & Nancarrow, B. E. (1999). Protest responses in contingent valuation. *Environmental and Resource Economics, 14*(1), 131–150.

Kahneman, D., Knetsch, J. L., & Thaler, R. H. (1990). Experimental tests of the endowment effect and the Coase theorem. *Journal of Political Economy, 98*(6), 1325–1348.

Karniouchina, E. V., Moore, W. L., van der Rhee, B., & Verma, R. (2009). Issues in the use of ratings-based versus choice-based conjoint analysis in operations management research. *European Journal of Operational Research, 197*(1), 340–348.

Kroes, E. P., & Sheldon, R. J. (1988). Stated preference methods. An introduction. *Journal of Transport Economics and Policy*, *22*(1), 11–25.

Louviere, J. J., & Hensher, D. A. (1982). On the design and analysis of simulated choice or allocation experiments in travel choice modelling. *Transportation Research Record, 890*(1), 11–17.

Louviere, J. J., Flynn, T. N., & Marley, A. A. J. (2015). *Best-worst scaling: Theory, methods and applications*. Cambridge University Press.

McFadden D. (1974). Conditional logit analysis of qualitative choice behavior. In P. Zarembka (Ed.), *Frontiers in econometrics* (pp. 104–142). Academic Press.

McFadden, D. (2017). *Stated preference methods and their applicability to environmental use and non-use valuations*. Edward Elgar Publishing.

Moore, W. L. (2004). A cross-validity comparison of rating-based and choice-based conjoint analysis models. *International Journal of Research in Marketing, 21*(3), 299–312.

Oglethorpe, D. R., & Miliadou, D. (2000). Economic valuation of the non-use attributes of a wetland: A case-study for Lake Kerkini. *Journal of Environmental Planning and Management, 43*(6), 755–767.

Payne, J. W., Bettman, J. R., & Johnson, E. J. (1992). Behavioral decision research: A constructive processing perspective. *Annual Review of Psychology, 43*(1), 87–131.

Rolfe, J., Bennett, J., & Louviere, J. (2000). Choice modelling and its potential application to tropical rainforest preservation. *Ecological Economics, 35*(2), 289–302.

Shogren, J. F., Shin, S. Y., & Hayes, D. J. (1994). Resolving differences in willingness to pay and willingness to accept. *The American Economic Review, 84*(1), 255–270.

Venkatachalam, L. (2004). The contingent valuation method: A review. *Environmental Impact Assessment Review, 24*(1), 89–124.

Walsh, R. G., Loomis, J. B., & Gillman, R. A. (1984). Valuing option, existence, and bequest demands for wilderness. *Land Economics, 60*(1), 14–29.

Whitehead, J. C., Pattanayak, S. K., Van Houtven, G. L., & Gelso, B. R. (2008). Combining revealed and stated preference data to estimate the non-market value of ecological services: An assessment of the state of the science. *Journal of Economic Surveys, 22*(5), 872–908.

Wilson, M. A., & Hoehn, J. P. (2006). Valuing environmental products and services using benefit transfer: The state-of-the art and science. *Ecological Economics, 60*(2), 335–342.

Chapter 3
The Fundamentals of Discrete Choice Experiment (DCE)

This chapter discusses the fundamental elements, logic, and key principles in designing DCE studies. Accordingly, this chapter will be rather *technical* as we cover the nuts and bolts of the DCE methodology in detail. As we will demonstrate in this chapter, a key strength of DCE is that it involves low cognitive complexity (i.e., easy to understand and respond to) for respondents when performing choice tasks. In a DCE, respondents are offered two or more alternatives in each choice task in which each alternative contains different combinations of attributes and attribute levels. They will then be asked to indicate which alternative they prefer in each choice task (as a dependent variable) in categorical format (choice A or B, or C). In the experiment, attributes and their levels are constructed so that respondents explicitly indicate their preference (thus the "stated preference" method). The last part of the DCE technique is the examination of the importance of each attribute for all respondents. In other words, DCE allows the estimation of the marginal rates of substitutions of the attributes (i.e., the monetary value of forgoing one attribute in favor of another attribute).

3.1 A Brief Overview of DCE

As we have discussed previously in Chap. 2, a discrete choice experiment (DCE) is a quantitative research methodology that assesses the relative importance of a set of products, services, ideas, or policy attributes that influence the *decision-making* of individuals such as consumers, entrepreneurs, investors, employees, parents, citizens, or voters. The DCE is designed to mimic the real-life decision-making process, where individuals are presented with different scenarios with varying attribute levels and

L. Shang and Y. Chandra, *Discrete Choice Experiments Using R*,
https://doi.org/10.1007/978-981-99-4562-7_3

asked to make a choice based on their preferences (Louviere et al., 2000). Each decision we make always involves multiple attributes (e.g., making a holiday decision involves making choices involving hotel brands, locations, service quality, prices, etc.). DCE is an attribute-based measure of preference. DCE is built on the assumptions that a situation (i.e., a product, a service, or an intervention) can be described by its attributes and that a decision-maker's valuation would depend upon different levels of these attributes. Attribute levels describe the variance over which attributes differ across alternatives. For example, when choosing a hotel for a holiday, a key attribute might be hotel brands, with three levels, such as Shangri-la vs. Hilton vs. Holiday Inn.

DCE, a form of quasi-experiment, is typically implemented in a *survey* format where respondents or participants are asked to *make choices* between a set of alternatives as defined by attributes and their corresponding levels. That is, respondents *state their choice* over different alternatives without requiring them to perform the alternatives (i.e., experiments without the actual intervention or quasi-experiment). As such, respondents provide quantitative information on their preference for each alternative, including trade-offs between attributes (Ryan & Gerard, 2003). DCE is based on the economic theory that humans or decision-makers will seek to *optimize their utility or welfare* (e.g., choosing Hilton over an unknown hotel brand for the same price per night) (Louviere et al., 2010). The data from the choice decisions made by respondents or participants allows researchers to uncover how respondents attach value to selected products, services, or intervention attributes.

Specifically, DCE provides insights into decision-makers' preferences and quantified information on the trade-offs they are willing to make between different product or service attributes and levels. The quantified information in DCE can be monetary or non-monetary valuation. DCE, among other stated preference methods, proves to be a valuable tool in situations where it is not feasible to use revealed preference data to assess the choices made by decision-makers, such as when valuing a new product that is still in the development phase and not yet accessible in the market, or when there is a lack of variation in available choices (Mangham & Hanson, 2008).

3.2 Theoretical Foundations of DCE

DCE draws upon two theories as its theoretical foundations. First, DCE follows Lancaster's *economic theory of value* (Lancaster, 1966), which argues that decision-makers have a preference for and derive utility from underlying *attributes* of products or services rather than the products or services per se. By presenting decision-makers with alternatives that vary in attributes and levels, DCEs provide researchers with insight into the preference structure of the respondents. This information is not readily available through other research methods. DCE helps elicit the *hidden* preference structure into an *observable* preference structure.

For example, it is difficult to know what people prefer from a box of chocolate for a Christmas gift. But we could scientifically describe the chocolate by flavor, calorie, packaging, and percentage of cocoa (using just three attributes at one level each as an example). Then, DCE can unravel what attributes of chocolate people value the most with high precision, transforming hidden preference structure into observable preference structure.

Secondly, DCE is rooted in the *random utility theory* (McFadden, 1974; Thurstone, 1927). Random utility theory is a long-standing theory of choice behavior that sees a decision-maker's behavior as an *interlinked* set of factors. The theory proposes that every decision-maker is a *rational* decision-maker who seeks to *maximize utility* relative to their choices and considers utility as a latent structure that cannot be observed directly. That is to say, random utility theory considers that a decision-maker has a "utility" for each choice alternative. Still, these utilities cannot be directly seen by researchers nor known precisely by the decision-maker.

Random utility theory originated from Thurstone's theory of paired comparisons (1927), in which Thurstone proposed that the modeling of individual choice is the outcome of a process of associating the random variable with each alternative and the alternative with the greatest realization of value is the one selected. Many years later, building on Thurstone's work, McFadden introduced the random utility framework as a tool in econometric analysis and assumed that each decision-maker faces a *finite choice set* and selects an alternative that maximizes their utility. In this framework, a choice is defined by the differences in attributes, marginal utilities for attributes, and a stochastic term (McFadden, 1973). The utility generated by an alternative is assumed to depend on the utility associated with its composing attributes and attribute levels.

Specifically, based on the random utility theory framework, the utility for individual i conditional on choice j can be separated into a *systematic* or *explainable* component V_{ij} and a *stochastic* or *non-explainable* component ε_{ij} (Mazzanti, 2003). The individual's utility U_{ij} for a particular alternative is represented as (Eq. 3.1):

$$\mathrm{U}_{ij} = V_{ij} + \varepsilon_{ij}, j = 1, \cdots, J \tag{3.1}$$

The non-explainable component ε_{ij} comprises all *unidentified factors* that impact individual choices from unobservable attributes, specification errors, and/or measurement errors.

The explainable component V_{ij} is a function of *attributes* of the product or service or idea or policy and *characteristics of individual decision-makers*, comprising *attributes* explaining differences in choice alternatives and *covariates* explaining differences in differences in individual choices, often represented as Eq. 3.2:

$$V_{ij} = X'_{ij}\beta + Z'_{i}\delta \tag{3.2}$$

where V_{ij} represents the utility or value associated with alternative j for individual i, X'_{ij} is a vector of *attributes of the alternatives* such as the hotel brand (e.g., Shangri-la), of good j as viewed by individual i, and Z' is a vector of *personal characteristics* for individual i, and β and δ are vectors of parameters representing *vectors of coefficients* to be estimated.

The utility U_{ij} depends on *attributes of the alternative j* and *attributes of the individual decision-maker i*. Random utility theory assumes that individuals will choose alternative j if, and only if, the utility derived from this alternative is *higher* than the utility of any other option in the set of J alternatives (Ben-Akiva et al., 1997; Louviere et al., 2010). Therefore, the probability P that individual i chooses option k from a set of competing alternatives C_i is represented as (Eq. 3.3):

$$\mathrm{P}(k|C_i) = \mathrm{P}[(V_{ik} + \varepsilon_{ik}) > \mathrm{Max}(V_{ij} + \varepsilon_{ij})], \tag{3.3}$$

for all options $j \neq k$ in the choice set C_i

3.3 The Key Elements of DCE Design

Before we explain the process of designing and implementing a DCE study, we first need to understand the key elements of DCE. As mentioned earlier, DCE is usually presented in a survey format where respondents are given a set of alternatives and asked to choose among different alternatives in a choice set and repeat the process for other choice sets. For example, suppose we want to study consumers' preference for bottled orange juice. In that case, we can present them with different choices of bottled orange juice described by a set of attributes and attribute levels. Tables 3.1 and 3.2 are examples of the potential choice sets (also known as a *choice question* in DCEs) that are presented to respondents in a typical DCE study. In Table 3.1, only two alternatives (Alternative A or B) are prompted for respondents to choose as the most preferred answer; while in Table 3.2, three alternatives (Alternative A or B or Neither) are used to allow respondents to choose "Neither" if both A or B are not considered attractive. The number of choices in each choice set is one of the key decisions researchers must make in the design stage.

The design stage in DCE is critical because it could influence how much utility information researchers can extract from respondents' answers (Louviere et al., 2000). A typical DCE design usually involves four key elements, which are *choice sets*, *alternatives*, *attributes*, and *attribute levels*:

- **Choice sets**: In a DCE study, respondents are presented with *multiple pairs* of alternatives for evaluation. A pair of alternatives is called a *choice set*. The examples in Tables 3.1 and 3.2 demonstrate a choice set in each study that examines respondents' preference toward orange juice and job choice, respectively. A DCE

Table 3.1 An example of a choice set in DCE (e.g., choosing orange juice)

Attributes	Alternative A 	Alternative B 
Organic production	Organic	Conventional
Sugar	15 g per 100 ml	25 g per 100 ml
Country of origin	Locally produced	Imported
Price	$5 per bottle	$8 per bottle
Which alternative do you prefer?		
Alternative A ☐		Alternative B ☐

Table 3.2 An example of a choice set in DCE (e.g., choosing jobs)

Attributes	Alternative A 	Alternative B 
Working hours	8 h per day	9 h per day
Location	Village	County
Facilities	Sufficient essential equipment	Nearly sufficient essential equipment
Training	Some	Sufficient
Monthly income	$3,000	$2,700
Which alternative do you prefer?		
Alternative A ☐	Alternative B ☐	Neither of these two ☐

study usually contains multiple choice sets that are presented and evaluated in multiple choice tasks by respondents.

- **Alternatives**: Alternatives are *choices* that are presented to respondents in a DCE study. Each alternative or scenario refers to a product or service that is described by a *set of attributes or characteristics*. In DCE, respondents are asked to *choose one alternative* they prefer more (or the most) out of a given number of alternatives (two or more) in *each choice task*. Table 3.1 provides two alternatives (A or B) for respondents to evaluate, while three alternatives (A or B or Neither) are exemplified in Table 3.2.
- **Attributes**: Attributes refer to *characteristics* of a product or service or idea or policy of interest to researchers in a DCE study. The attributes are used to describe *different characteristics of the alternatives*. From the choices made by respondents in a DCE study, researchers can elicit the *relative importance* that respondents

place on the attributes of a product or service and how they are willing to make trade-offs of one attribute over another. In Table 3.1, which examines decision-makers' preference for orange juice, four attributes are included to describe orange juice: *Organic Production, Sugar, Country of Origin, and Price.*

- **Attribute levels**: Attributes levels refer to *values assigned to each attribute.* We can think of attribute level as the operationalization of each attribute. The attribute levels need to be constructed so that the respondents are willing to make trade-offs (i.e., sacrificing one alternative in favor of another) between combinations of different attributes. For example, the attribute Country of Origin in Table 3.1 has two attribute levels, including *locally produced and imported.* This will consciously ask respondents to state their choice of the country of origin of the orange juice.

3.4 Designing and Implementing DCE

The process of designing and implementing DCE can be divided into six steps, which are adapted from the model proposed by Ryan (1996). A simple process flowchart is illustrated below (see Fig. 3.1).

In designing a DCE experiment, a researcher needs to first establish a clear understanding of the issue (e.g., research question or unresolved theoretical puzzle and the research objective) for examination. For instance, if your study aims to understand patients' preference for flu vaccination, then the DCE should be designed and constructed around factors that may influence people's decisions about whether to receive a vaccination. The researcher should justify the utilization of a DCE design over other approaches (i.e., limited vaccination alternatives available; lack of access to patients' records; new vaccines available to be tested on humans).

Step 1: Identification of Attributes: The first step in designing a DCE study involves *identifying the attributes* of interest that will be used to describe the product or service or idea or policy under investigation. Selecting and defining attributes require a good understanding of the topic, literature, and the target population's perspective (Coast & Horrocks, 2007). There are multiple ways of identifying attributes, such as conducting *systematic reviews* to identify potential attributes to be used in the study and collecting primary data via in-depth interviews or focus group discussions with potential respondents or participants. The selected attributes should be *demand relevant* (i.e., something consumers are willing to pay a price for) *and measurable.* Normally, a DCE study consists of no more than ten attributes (i.e., usually up to 7 attributes) for ease of operationalization and to avoid respondent fatigue in completing the study.

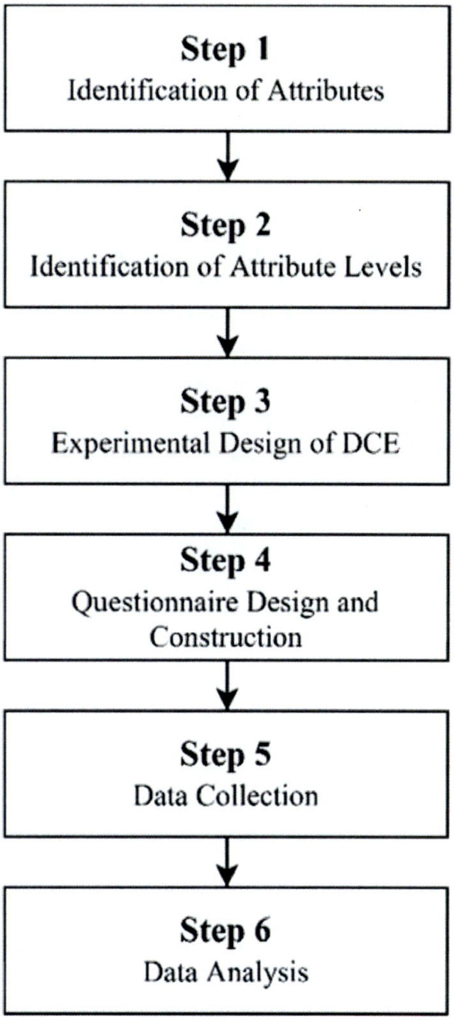

Fig. 3.1 Stages of a DCE study

Step 2: Identification of Attribute Levels: After deciding which attributes to use to describe the product, service, idea, or policy for evaluation, the next step is to decide the values associated with each attribute *or the attribute levels*. To determine attribute levels, researchers could consider different situations that potential decision-makers may encounter. Typically, the attribute levels selected should reflect the *most plausible range* of situations that target populations are expected to experience. The levels could be either qualitative categories (e.g., locally produced vs. imported for Country of Origin attribute) or quantitative measures ($10, $15, or $20 for Price attribute) (see Table 3.1). The attribute levels should be actionable to the respondents in a way that they are willing to make trade-offs between combinations of the attributes. Researchers can do in-depth interviews and or focus group discussions as part of a due diligence process to ensure that the attributes and attribute levels are appropriate in the study.

Step 3: Experimental Design of DCE: In this step, researchers design and construct the *alternatives* or *choice sets* based on the attributes and attribute levels identified earlier. Good experimental design is crucial for the success of a DCE experiment. Researchers will make a series of decisions in designing a DCE experimental design. Examples include deciding between using a *full factorial design* or a *fractional factorial design*, the number of alternatives in each choice set, the allocation of the alternatives, the *number of choice sets* to present in each choice task to respondents, and the number of respondents to study. We will provide step-by-step instructions with examples to illustrate how to design a DCE experiment using R in Chap. 7.

Step 4: Questionnaire Design and Construction: The next step is to develop a questionnaire that will be distributed to respondents or participants for primary data collection. Usually, a DCE questionnaire includes these sections: (1) information on the objectives of the study and an explanation of the DCE task, (2) screening questions based on certain (inclusion or exclusion) criteria, (3) the DCE choice sets, (4) demographic questions, and (5) follow-up questions to examine respondents' preference further. In this step, researchers should consider whether the task instructions are appropriate and easy to understand and whether sufficient background and contextual information are provided to respondents. Doing a pre-test of the DCE questionnaire prior to the actual launch is advisable to ensure the wording and definition of attributes and attribute levels and that participants fully understand choice tasks. Researchers can take advantage of software such as Qualtrics, Google Survey, Survey Monkey, or Question Pro to assist in implementing the DCE survey. Notably, after completing the questionnaire design and before data collection, it is important for researchers to ensure that they obtain ethics approval as a DCE involves human participants.

Step 5: Data Collection: After the finalization of the DCE questionnaire, researchers need to decide on the methods to collect DCE data. Common methods used in DCE data collection include *face-to-face interviews*, *telephone interviews*, *mailed paper questionnaires*, *online questionnaires*, or combining multiple methods (e.g., qualitative interviews followed by an online DCE survey). The advantage of using an online questionnaire in DCE for data collection is that researchers can use the randomization process enabled by survey software for DCE data collection. In certain DCE designs, randomization is needed because the experimental design is too complex to do. Thus, the DCE study is divided into multiple blocks design, and each respondent is randomly assigned to only one of the block designs. R statistical programming offers a powerful suite of packages to assist in the optimal design of choice sets or fractional/full factorial design.

Step 6: Data Analysis: The final step in DCE is data analysis, reporting, and result interpretation. Generally, the analysis of DCE data involves logit-based regression models with a dichotomous or polychotomous categorical dependent variable (i.e., the dependent variable in DCE is usually a choice of alternative A, B, or C). When the choice presented to respondents is binary (a single yes or no vote on an alternative) or comprises two alternatives (i.e., alternative A or B), *binary probit or logit* (logistic regression) models are more appropriate. When "Neither" or "Opt-out" alternatives are included in the choice set, or there are more than two alternatives, McFadden's *multinomial logit* (MNL) regression is more suitable (McFadden, 1974). We will discuss the data analysis process of DCE in more detail in Chap. 9 later.

Figure 3.1 only provides a simplified overview of the key steps involved in conducting a DCE. However, it should be noted that some of these tasks are not strictly sequential but rather inter-related and may influence each other. For example, the experimental design is also closely linked to the identification of attributes and attribute levels. The selection of attributes and their levels is a crucial step in a DCE, as these determine the factors that will be varied in the choice sets presented to participants. The experimental design aims to create a well-balanced and efficient set of choice sets that allow for accurate estimation of preferences and meaningful analysis of the data. The choice of attribute levels and the arrangement of choice sets should be guided by statistical principles, such as orthogonality and efficiency, to optimize the information obtained from participants.

3.5 Various Applications of DCE

DCE experiments have been widely applied in different disciplines. It has become one of the most applied choice modeling methods. In the tables below, we summarized several empirical studies that used the DCE method in transportation economics, healthcare, tourism management, and environmental economics. We highlighted the topics of investigation, selected attributes, number of respondents, and number of alternatives in each choice set in Table 3.3. These will offer benchmarks for any researchers wishing to use DCE.

Table 3.3 Sample DCE applications in different disciplines

Research discipline	Sample areas of interest	Sample articles	DCE topics	Attributes used	Attribute levels	Experimental design	Sample size	Alternatives in each choice set
Transportation								
Transportation	City transport system	Gundlach et al. (2018)	Travelers' preferences for car-free city centers	(1) Road network for cyclists (2) Walking distance to closest public transport stops (3) Frequency of public transport (4) Park and ride facilities at public transport stops bordering the car-free city center (5) Additional recreation areas (6) Price for public transport	(1) As today vs Bikeways next to every road vs Separate road network (2) As today vs 6 min vs 3 min (3) As today vs More frequent vs Much more frequent (4) As today vs Unguarded parking lot vs Guarded parking lot (5) As today vs 20% more vs 40% more (6) Free of charge vs 75% less vs 50% less vs 25% less vs As today vs 25% more	Blocked fractional factorial design (i.e., 27 choice sets split into three blocks)	334	Two alternatives and one opt-out option

(continued)

Table 3.3 (continued)

Research discipline	Sample areas of interest	Sample articles	DCE topics	Attributes used	Attribute levels	Experimental design	Sample size	Alternatives in each choice set
		König and Grippenkoven (2020)	Travelers' appraisal of ridepooling services	(1) Time of booking (2) Walking distance (3) Shift of departure (4) Travel time (5) Information provision (6) Fare	(1) 5 min vs 10 min vs 30 min (2) 0 vs 300 m vs 500 m (3) 0 vs 10 min vs 20 min (4) 10 min vs 20 min vs 30 min (5) None vs Few vs Much (6) €2.5 vs €3 vs €3.5 vs €4	Fractional factorial design (i.e., 24 choice sets in total for each respondent)	410	Two alternatives and one opt-out option
	Transport behaviors	McNamara et al. (2013)	Older people's preferences for relinquishing their driver licenses	(1) Percentage risk of car crash in the next year (2) Age (3) Confidence levels in your own diving ability (4) Recommendations by others about your fitness to drive (5) Availability of other transport options (6) Cost of public transport for older people	(1) 5% vs 30% 60% (2) 70 vs 80 vs 90 (3) Highly confident vs Medium confident vs Low confident (4) Local doctor recommends you stop driving vs Local doctor says you are fit to drive but your family and friends recommend that you stop driving vs Both local doctor and family and friends recomend that you are fit to drive (5) Available to you all of the time vs Available to you some of the time vs Hardly ever available to you (6) Free public transport at all times vs Free public transport 9 a.m. to 3 p.m. week days only vs 25% consession off the full fare at all times	Blocked fractional factorial design (i.e., 18 choice sets split into three blocks)	114	Two alternatives
		Van Acker et al. (2020)	Preferences for long-distance coach transport	(1) Travel time (2) Wi-Fi (3) Leg space (4) Catering (5) Entertainment (6) Power plug (7) Travel cost/ price	(1) Slow vs Average vs Fast (2) No vs Yes (3) Standard vs Luxurious (4) None vs Snacks vs Hot meal (5) Collective vs Individual (6) No vs Yes (7) Expensive vs Average vs Cheap	Fractional factorial design (i.e., 16 choice sets in total for each respondent)	274	Two alternatives

(continued)

Table 3.3 (continued)

Research discipline	Sample areas of interest	Sample articles	DCE topics	Attributes used	Attribute levels	Experimental design	No. of respondents	No. of alternatives in each choice set
Healthcare								
Healthcare	Treatment/ care alternatives	Shah et al. (2015)	Public's preferences regarding life-extending, end-of-life treatments	(1) Life expectancy without treatment (2) Qualify-of-life gain from treatment (3) Life expectancy gain from treatment (4) Qualify-of-life without treatment	(1) 3 vs 12 vs 24 vs 36 vs 60 months (2) 50% vs 100% (3) 0 vs 1 vs 2 vs 3 vs 6 vs 12 months (4) 0% vs 25% vs 50%	Blocked fractional factorial design (i.e., 80 choice sets split into eight blocks)	3969	Two alternatives and one opt-out option
		Morton et al. (2012)	Patients' preferences for dialysis and conservative care to treat end-stage kidney disease	(1) Average life expectancy (2) Visits to hospital for dialysis (3) Ability to travel or take short trips (4) Time spent undergoing dialysis (5) Time of day dialysis can be done (6) Subsidized transport provided for attending treatment or appointments (7) Ability to change day or time of dialys	(1) 2 vs 5 vs 10 vs 15 years (2) 0 vs 2 vs 3 vs 4 visits (3) Not restricted vs Slightly restricted vs Very restricted (4) 0 vs 4 vs 8 vs 10 h (5) Day vs Day or evening vs Night vs Not applicable (6) Not provided vs Provided at small cost to me vs Provided at no cost to me (7) Not applicable vs Up to once per month vs Up to once per week vs Whenever necessary	Fractional factorial design (12 choice sets for each respondent)	105	Two alternatives and one opt-out option

(continued)

Table 3.3 (continued)

Research discipline	Sample areas of interest	Sample articles	DCE topics	Attributes used	Attribute levels	Experimental design	No. of respondents	No. of alternatives in each choice set
	Accessibility and quality of health services	Rowen et al. (2022)	Healthcare stakeholders' preferences for aspects of mental healthcare quality and accessibility	(1) Waiting times (2) Ease of access (3) Person-centred care (4) Co-ordinated approach (5) Continuity (6) Communication, capacity and resources (7) Treated as a person (8) Recovery focus (9) Inappropriate discharge (10) Quality of life	(1) Yes vs No (2) Yes vs No (3) Yes vs No (4) Yes vs No (5) Yes vs No (6) Yes vs No (7) Yes vs No (8) Yes vs No (9) Yes vs No (10) 20% vs 50% vs 90% quality of life	Blocked fractional factorial design (i.e., 32 choice sets split into four blocks)	1859	Two alternatives
		Hanson et al. (2005)	Preferences for hospital quality	(1) Likelihood that the child will receive all the drugs (2) Likelihood that the hospital staff will examine the child properly (3) Attitudes of staff (4) Cleanliness of wards and toilets (5) Waiting time between arrival at OPD (6) Cost	(1) Drugs available vs Likely to have to seek some drugs elsewhere (2) Child examined thoroughly vs Child given a superficial examination (3) Staff are indifferent vs Staff are friendly vs Staff are rude (4) Cleaned rarely vs Cleaned often (5) 5 h vs 2 h vs 30 min (6) K100000 vs K20000 vs K250000	Fractional factorial design (16 choice sets for each respondent)	2308	Two alternatives

(continued)

Table 3.3 (continued)

Research discipline	Sample areas of interest	Sample articles	Publishers	DCE topics	Attributes used	Attribute levels	Experimental design	No. of respondents	No. of alternatives in each choice set
Tourism									
Tourism	Preferences for leisure recreation	De Valck et al. (2017)	Land Use Policy	Consumers' preferences for outdoor recreation destinations	(1) Distance (2) Facilities (3) Signed trail quality (4) Tranquillity (5) Presence of a water element (6) Landscape openness (7) Landscape naturalness (8) Landscape diversit	(1) Status quo vs Move away from the status quo (2) Large supply of recreational facilities vs No recreational facilities vs Limited supply of recreational facilities (3) Mostly covered with paved signed trails vs Absence of any signed trails vs Mostly covered with unpaved signed trails (4) Low vs High vs Intermediate noise level (5) At least one pond or lake vs No water element vs At least one river or canal (6) Open vs Closed vs Semi-open landscape (7) Natural vs Rural vs Semi-natural landscape (8) Diverse vs Uniform vs Semi-diverse landscap	Blocked fractional factorial design (i.e., 36 choice sets split into six blocks)	1403	Two alternatives and one opt-out option

(continued)

Table 3.3 (continued)

Research discipline	Sample areas of interest	Sample articles	Publishers	DCE topics	Attributes used	Attribute levels	Experimental design	No. of respondents	No. of alternatives in each choice set
		Shoji and Tsuge (2015)	Tourism Economics	Visitors' preferences for winter nature-based tours	(1) Purpose of the tour (2) Interpretation (3) Possibility of finding eagles (4) Fee for a tour	(1) Cross-country skiing or snowshoeing vs Drift ice vs Wildlife observation (2) Detailed interpretation vs No detailed interpretation (3) 20% vs 50% vs 90% per tour (4) 2000 vs 4000 vs 6000 vs 8000 vs 10,000 JPY	Fractional factorial design (six choice sets for each respondent)	116	Two alternatives and one opt-out option
	Accommondation choices	Kim and Park (2017)	Toursim Management	Hotel consumers' preferences for different aspects of hotel choices	(1) Price (2) Service & Food quality (3) National, recognized brand (4) Sports facilities (5) Comfortable (6) Entertaining (7) Room quality (8) Overall atmosphere	(1) $ vs $$$ vs $$$$$ (2) 1 Star vs 3 Star vs 5 Star (3) Yes vs No (4) 3 Star vs 5 Star (5) 1 Star vs 3 Star vs 5 Star (6) 1 Star vs 3 Star vs 5 Star (7) 1 Star vs 3 Star vs 5 Star (8) 1 Star vs 3 Star vs 5 Sta	D-optimal design (each respondent is randomly assigned to one of the ten different survey designs, each with 16 choice sets)	494	Two alternatives and one opt-out option

(continued)

Table 3.3 (continued)

Research discipline	Sample areas of interest	Sample articles	Publishers	DCE topics	Attributes used	Attribute levels	Experimental design	No. of respondents	No. of alternatives in each choice set
		Masiero et al. (2015)	International Journal of Hospitality Management	Tourists' preferences for hotel rooms	(1) View (2) Floor (3) Access to hotel club (4) Free mini bar (5) Guest smartphone (6) Cancellation (7) Price per room per night	(1) City vs Harbour (2) 10th vs 18th vs 26th (3) No vs Yes (4) Soft drinks, snacks vs Soft drinks, snacks, wine & beer (5) Not available vs Available (6) Non-refundable vs Refundable (7) 1600 vs 2000 vs 2400 vs 2800 vs 3200	Fractional factorial design (12 choice sets for each respondent)	808	Two alternatives

Research discipline	Sample areas of interest	Sample articles	DCE topics	Attributes used	Attribute levels	Experimental design	No. of respondents	No. of alternatives in each choice set
Environment								
Environment	Preferences for nature conservation	De Valck et al. (2014)	Public's preferences and willingness to pay for nature restoration	(1) Habitat (2) Reduction in coniferous forest (3) Biodiversity (4) Accessibility (5) Price	(1) Conversion from coniferous forest to heathland vs Conversion from coniferous forest to broadleaved forest (2) More species (common species) vs More species (common and rare species) (3) 50 ha vs 100 ha vs 200 ha (4) Good vs Poor (5) 10 vs 25 vs 50 vs 75 vs 125 vs 200	Blocked fractional factorial design (i.e., 24 choice sets split into four blocks)	217	Two alternatives and one opt-out option

(continued)

Table 3.3 (continued)

Research discipline	Sample areas of interest	Sample articles	DCE topics	Attributes used	Attribute levels	Experimental design	No. of respondents	No. of alternatives in each choice set
		Notaro and Grilli (2022)	Tourists' preferences for conservation of carnivores	(1) Wolves (2) Lynx (3) Salamanders (4) Cost	(1) 0 vs 15 vs 30 vs 45 vs 60 vs 75 vs 90 (2) 0 vs 15 vs 30 vs 45 vs 60 vs 75 vs 90 (3) 0 vs 15 vs 30 vs 45 vs 60 vs 75 vs 90 (4) 0 vs 3 vs 6 vs 9 vs 12 vs 15 vs 18	Bayesian efficient design (12 choice sets for each respondent)	420	Two alternatives and one opt-out option
	Environmental protection and controls	Ndunda and Mungatana (2013)	Preferences for improved wastewater treatment	(1) Quality of wastewater for irrigation (2) Quantity of wastewater for irrigation (3) Ecosystem restoration in Motoine–Ngong–Nairobi River (4) Monthly municipal tax	(1) Poor vs Medium vs High (2) Low vs Medium vs High (3) No vs Yes (4) 60 vs 120 vs 160 vs 200 vs 24	Blocked fractional factorial design (i.e., 64 choice sets split into eight blocks)	241	Two alternatives and one opt-out option
		Tang and Zhang (2016)	Public's preferences for air quality improvement	(1) Days (2) Mortality (3) Policy (4) Delay (5) Cost	(1) Day 15 vs Day 20 vs Day 25 (2) 180 k vs 270 k vs 350 k (3) Limit private transportation vs Update factory pollutant disposal (4) 1 yr vs 5 yrs vs 10 yrs (5) 1000 vs 2000 vs 3000 RMB	Fractional factorial design (18 choice sets for each respondent)	200	Two alternatives

References

Ben-Akiva, M., McFadden, D., Abe, M., Böckenholt, U., Bolduc, D., Gopinath, D., & Steinberg, D. (1997). Modeling methods for discrete choice analysis. *Marketing Letters, 8*(3), 273–286.

Coast, J., & Horrocks, S. (2007). Developing attributes and levels for discrete choice experiments using qualitative methods. *Journal of Health Services Research & Policy, 12*(1), 25–30.

De Valck, J., Vlaeminck, P., Broekx, S., Liekens, I., Aertsens, J., Chen, W., & Vranken, L. (2014). Benefits of clearing forest plantations to restore nature? Evidence from a discrete choice experiment in Flanders, Belgium. *Landscape and Urban Planning, 125*, 65–75.

De Valck, J., Landuyt, D., Broekx, S., Liekens, I., De Nocker, L., & Vranken, L. (2017). Outdoor recreation in various landscapes: Which site characteristics really matter? *Land Use Policy, 65*, 186–197.

Garrod, G., & Willis, K. G. (1999). Economic valuation of the environment. *Books*.

Gundlach, A., Ehrlinspiel, M., Kirsch, S., Koschker, A., & Sagebiel, J. (2018). Investigating people's preference for car-free city centers: A discrete choice experiment. *Transportation Research Part D: Transport and Environment, 63*, 677–688.

Hanson, K., McPake, B., Nakamba, P., & Archard, L. (2005). Preferences for hospital quality in Zambia: Results from a discrete choice experiment. *Health Economics, 14*(7), 687–701.

Kim, D., & Park, B. J. R. (2017). The moderating role of context in the effects of choice attributes on hotel choice: A discrete choice experiment. *Tourism Management, 63*, 439–451.

Kolstad, J. R. (2011). How to make rural jobs more attractive to health workers. Findings from a discrete choice experiment in Tanzania. *Health Economics, 20*(2), 196–211.

König, A., & Grippenkoven, J. (2020). Modelling travelers' appraisal of ridepooling service characteristics with a discrete choice experiment. *European Transport Research Review, 12*(1), 1–11.

Lancaster, K. J. (1966). A new approach to consumer theory. *Journal of Political Economy, 74*(2), 132–157.

Louviere, J. J., Hensher, D. A., & Swait, J. D. (2000). *Stated choice methods: Analysis and applications*. Cambridge University Press.

Louviere, J. J., Flynn, T. N., & Carson, R. T. (2010). Discrete choice experiments are not conjoint analysis. *Journal of Choice Modelling, 3*(3), 57–72.

Mangham, L. J., & Hanson, K. (2008). Employment preference of public sector nurses in Malawi: Results from a discrete choice experiment. *Tropical Medicine & International Health, 13*(12), 1433–1441.

Masiero, L., Heo, C. Y., & Pan, B. (2015). Determining guests' willingness to pay for hotel room attributes with a discrete choice model. *International Journal of Hospitality Management, 49*, 117–124.

Mazzanti, M. (2003). Discrete choice models and valuation experiments. *Journal of Economic Studies*.

McFadden, D. (1973). Conditional logit analysis of qualitative choice behavior.

McFadden, D. (1974). The measurement of urban travel demand. *Journal of Public Economics, 3*(4), 303–328.

McNamara, A., Chen, G., George, S., Walker, R., & Ratcliffe, J. (2013). What factors influence older people in the decision to relinquish their driver's licence? A discrete choice experiment. *Accident Analysis & Prevention, 55*, 178–184.

Monchambert, G. (2020). Why do (or don't) people carpool for long distance trips? A discrete choice experiment in France. *Transportation Research Part A: Policy and Practice, 132*, 911–931.

Morton, R. L., Snelling, P., Webster, A. C., Rose, J., Masterson, R., Johnson, D. W., & Howard, K. (2012). Factors influencing patient choice of dialysis versus conservative care to treat end-stage kidney disease. *CMAJ, 184*(5), E277–E283.

Ndunda, E. N., & Mungatana, E. D. (2013). Evaluating the welfare effects of improved wastewater treatment using a discrete choice experiment. *Journal of Environmental Management, 123*, 49–57.

Notaro, S., & Grilli, G. (2021). Assessing tourists' preference for conservation of large carnivores in the Italian Alps using a discrete choice experiment. *Journal of Environmental Planning and Management*, 1–20.

Rockers, P. C., Jaskiewicz, W., Wurts, L., Kruk, M. E., Mgomella, G. S., Ntalazi, F., & Tulenko, K. (2012). Preference for working in rural clinics among trainee health professionals in Uganda: A discrete choice experiment. *BMC Health Services Research, 12*(1), 1–13.

Rowen, D., Powell, P. A., Hole, A. R., Aragon, M. J., Castelli, A., & Jacobs, R. (2022). Valuing quality in mental healthcare: A discrete choice experiment eliciting preference from mental healthcare service users, mental healthcare professionals and the general population. *Social Science & Medicine, 301*, 114885.

Ryan, M. (1996). Using willingness to pay to assess the benefits of assisted reproductive techniques. *Health Economics, 5*(6), 543–558.

Ryan, M., & Gerard, K. (2003). Using discrete choice experiments to value health care programmes: Current practice and. *Applied Health Economics and Health Policy, 2*(1), 55–64.

Shah, K. K., Tsuchiya, A., & Wailoo, A. J. (2015). Valuing health at the end of life: A stated preference discrete choice experiment. *Social Science & Medicine, 124*, 48–56.

Shoji, Y., & Tsuge, T. (2015). Heterogeneous preference for winter nature-based tours in sub-frigid climate zones: A latent class approach. *Tourism Economics, 21*(2), 387–407.

Thurstone, L. L. (1927). A law of comparative judgment. *Psychological Review, 34*(4), 273.

Zha, D., Yang, G., Wang, W., Wang, Q., & Zhou, D. (2020). Appliance energy labels and consumer heterogeneity: A latent class approach based on a discrete choice experiment in China. *Energy Economics, 90*, 104839.

Chapter 4
A Review of R and Its Applicability to DCE

In this chapter, we review the role of R as a computing platform and introduce various R packages to support the construction and implementation of DCE experiments. We discuss the benefits of using R and R packages as open-source software to support DCE studies. We also provide brief examples of installing and operating R for new R users.

4.1 An Overview of R

R (http://www.r-project.org) is an open-source programming language widely utilized as a statistical software and data analysis tool within the GNU package. R is a popular tool to do “data science”. R compiles and runs on various operating systems, including UNIX platforms and similar systems, Windows, and MacOS. R uses command-line scripting, and users can instruct R what to do and interact with R in a (default) console window where R will also show the command results immediately. A screenshot of the R console running under its default GUI interface in R version 4.2.0[1] in MS Windows is shown below in Fig. 4.1.

R provides a wide variety of statistical and graphical capabilities, from basic to advanced. R is highly extensible by allowing users to freely distribute, study, change, and improve the software. R user communities have created packages to augment the functions of the R language. R is a versatile tool that can be used for a wide range of tasks, including *data manipulation* (i.e., shaping the dataset into a desired format that could be easily used and analyzed later), *analysis*, and *visualization*. With its vast collection of over 18,000 packages on the Comprehensive R Archive

[1] By Feb 2023, the latest stable release of R version is 4.2.2. To get more information about R and its latest updates, you can visit the official R website at https://www.r-project.org/.

L. Shang and Y. Chandra, *Discrete Choice Experiments Using R*,
https://doi.org/10.1007/978-981-99-4562-7_4

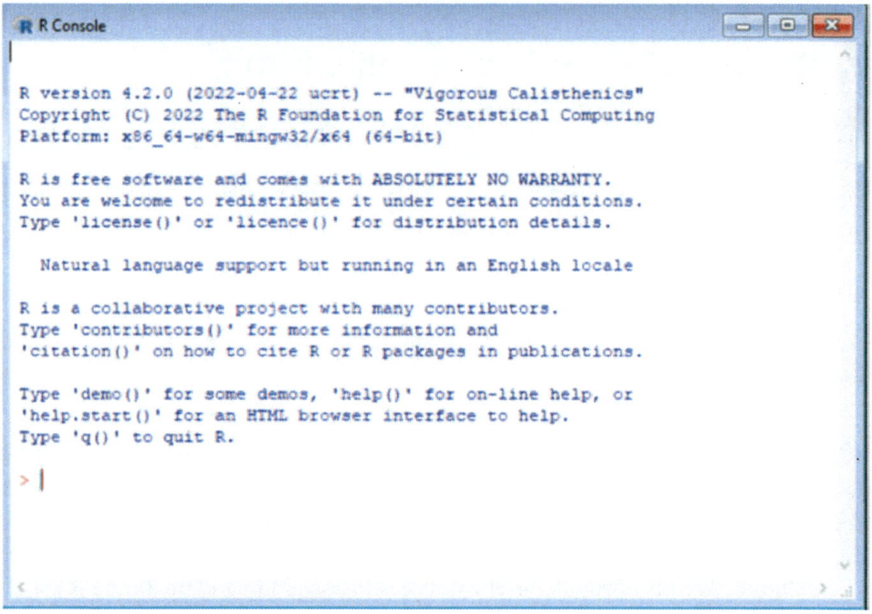

Fig. 4.1 A screenshot of the R console

Network (CRAN) as of May 2022,[2] R provides a wealth of options for statistical analysis. Additionally, R includes a variety of built-in functions and packages for creating high-quality visualizations, making it a powerful tool for exploring and communicating insights from data. For researchers, R could be a free but powerful alternative to traditional statistical packages such as SAS, SPSS, and Stata.

4.2 Installation of R and RStudio

Because Windows OS is mostly available to computer users, we use Windows for the following demonstration. To install R on Windows, users should go to the CRAN website[3] and download the R installer. After clicking the link, you will be directed to a page (see Fig. 4.2) that shows the download link of the most updated version of R (v4.2.0 as of May 2022).[4]

Once the installer downloads, you may run this program and follow the setup instructions shown in the installation wizard. The wizard will install R into your *program file* folders (or any other folders you manually select) and create a *shortcut* in your Start menu.

---

[2] https://cran.r-project.org/web/packages/.

[3] https://cran.rstudio.com/bin/windows/base/.

[4] The page display and content may be replaced by the most current version of R.

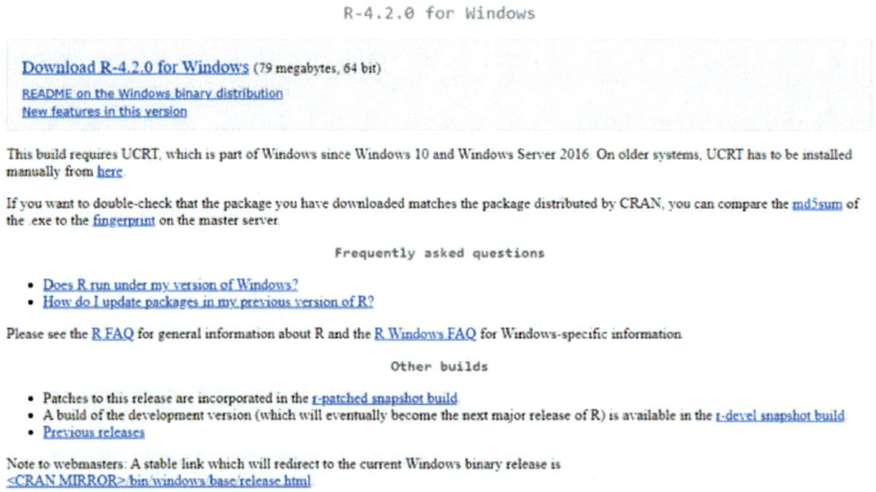
R-4.2.0 for Windows

Download R-4.2.0 for Windows (79 megabytes, 64 bit)
README on the Windows binary distribution
New features in this version

This build requires UCRT, which is part of Windows since Windows 10 and Windows Server 2016. On older systems, UCRT has to be installed manually from here.

If you want to double-check that the package you have downloaded matches the package distributed by CRAN, you can compare the md5sum of the .exe to the fingerprint on the master server.

Frequently asked questions

- Does R run under my version of Windows?
- How do I update packages in my previous version of R?

Please see the R FAQ for general information about R and the R Windows FAQ for Windows-specific information.

Other builds

- Patches to this release are incorporated in the r-patched snapshot build.
- A build of the development version (which will eventually become the next major release of R) is available in the r-devel snapshot build.
- Previous releases

Note to webmasters: A stable link which will redirect to the current Windows binary release is
<CRAN MIRROR>/bin/windows/base/release.html

Fig. 4.2 R download page for windows users

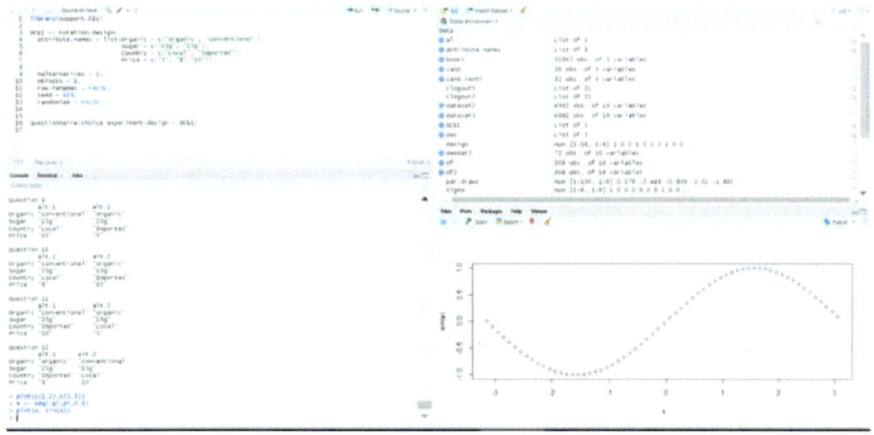

Fig. 4.3 A docked windows in RStudio

In addition to installing R, we suggest users download and install RStudio, an integrated development environment for R. RStudio provides an elegant interface that improves user experience with the R software (RStudio Team, 2021). First, RStudio allows users to dock and arrange relevant windows on the screen so that users can keep the code, images, comments, and plots *all together in a screen display*, as shown in Fig. 4.3.

Secondly, RStudio allows users to write syntax, scripts, or codes in a markdown format instead of in a normal text file. R Markdown (as shown in the upper left pane of RStudio in Fig. 4.3) allows users to write code, text, and visualizations in a single document, making it easier to organize and share the results of data analyses. With this function, you could seamlessly document your codes together with comments and/or notes and see both the code and results separately in the console after running the code.

Furthermore, the RStudio console (shown in Fig. 4.4) allows users to interactively run and test code, view outputs, and access the history of commands used. It also provides access to various tools and features such as debugging, profiling, and package installation. The console is an essential component of the RStudio interface, providing users with a direct and immediate way to interact with R and execute code.

Furthermore, RStudio has an autocomplete feature that suggests potential code while the user is writing code by using the tab key. This function can save time and reduce typos, which is particularly helpful for new R users. For instance, if a user types "*install*" and presses the tab key, RStudio will suggest completing the code with "*install.packages*", as shown in Fig. 4.5.

To install RStudio, you must install the R-base first, or else RStudio will not work. From the RStudio website, go to the download page (https://www.rstudio.com/products/rstudio/download/) and click the *download* button for RStudio Desktop. After downloading the RStudio installer, you may follow the installation instructions for further setup.

Overall, installing R and RStudio is quite easy as many instructions and user manuals are available online that offer step-by-step guidance. For R and RStudio installation in other operating systems (e.g., Linux, MacOS), go to this webpage

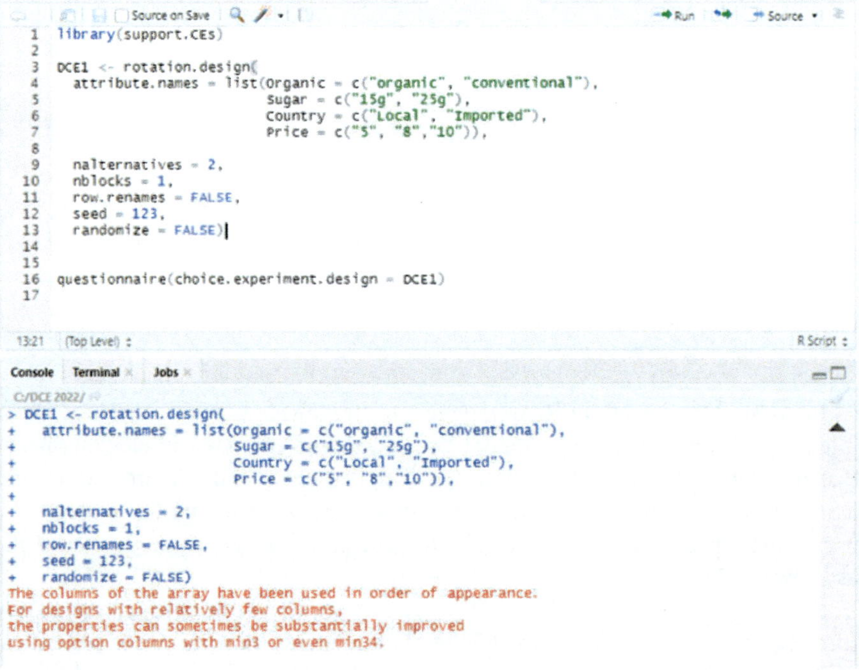

Fig. 4.4 An RStudio console

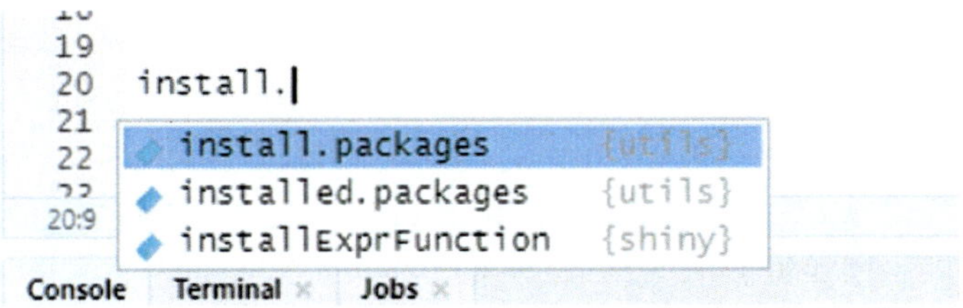

Fig. 4.5 An autocompleting syntax in RStudio

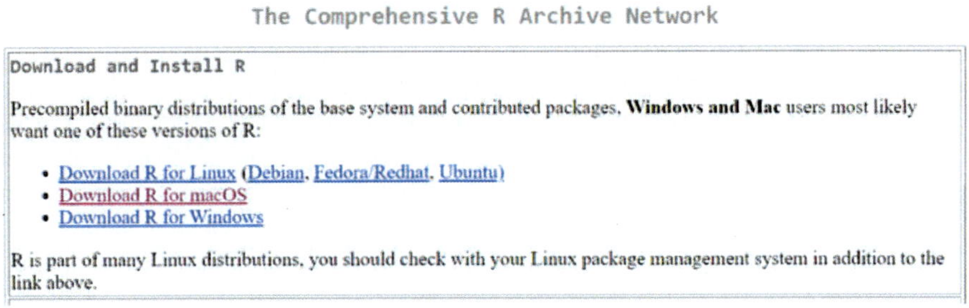

Fig. 4.6 Downloading R for various operating systems

http://lib.stat.cmu.edu/R/CRAN/ and find the corresponding operating system for more installation instructions (see Fig. 4.6).

4.3 Installing R Packages

In the R language, the fundamental unit of shareable code is the R *package* (*e.g., ggplot2* for creating complex data plots, *dplyr* for data manipulation, *lm* for linear regression). An R package is stored under a *library* of packages that have been developed and maintained by R developers to solve some problems and/or cover some needs that are not met in base R. R package bundles together code, data, documentation, and tests, and can be easily shared with other users. We need to rely on *various R packages* to conduct a DCE study using R. Before we discuss and demonstrate the packages to conduct DCE studies, we will first show you how to *install* and *implement packages* in R.

There are several ways to install R packages. First, you could call the *install.package* function in R or RStudio Console with the package name enclosed in quotation marks inside the parentheses. For example, if we want to install a package called **support.CEs** that supports the experimental design of DCE studies, we can use the following code for installation (refer to Fig. 4.7 for an example of installing R packages by using the code). It is important to note that R language is **case sensitive** (e.g., *Support.CEs* or *support.ces* is incorrect but *support.CEs* is correct) and

spacing sensitive (e.g., *support. CEs* is incorrect but *support.CEs* is correct). Thus, precision is everything in R coding.

install.packages("support.CEs")

Alternatively, you can also choose to install packages from the menu function. To install a package using the RStudio menu function, you can follow these steps (see Fig. 4.8):

1. Open RStudio and navigate to the Tools menu.
2. Select "Install Packages" from the drop-down menu.
3. Select "Install from Repository (CRAN)" in the pop-up window.
4. In the "Packages" field, enter the name(s) of the package(s) you wish to install, separated by commas if installing multiple packages.
5. Optionally, you can choose a specific repository from the "Repositories" drop-down menu, although the default CRAN repository should work for most users.
6. Click "Install" to start the installation process.

Once a package is installed on your system, it should remain available for use until you uninstall it or update it to a newer version.

Similarly, if you are using R Console, go to Packages, click on Install package(s)..., and select a CRAN mirror in the pop-up window (you may select the location nearest to you to minimize network load and time for installation). After this, you will see a new window with a list of CRAN packages available for you to choose from for installation (see Fig. 4.9).

After installation, you need to *load the R package* from the library to *access its functions*. To do this, you can use the ***library()*** function, specifying the package name as below:

library(support.CEs)

The *library ()* function will make the functions and capabilities of the specified package available for use in your R code. Once an R package is successfully loaded, you can use the ***help()*** function with the package name to access its documentation that contains instructions and useful examples of how the package works. The screenshot below (see Fig. 4.10) shows an example of the documentation of the package **support.CEs** in RStudio. Basically, it provides a description, details, and examples of how the package works.

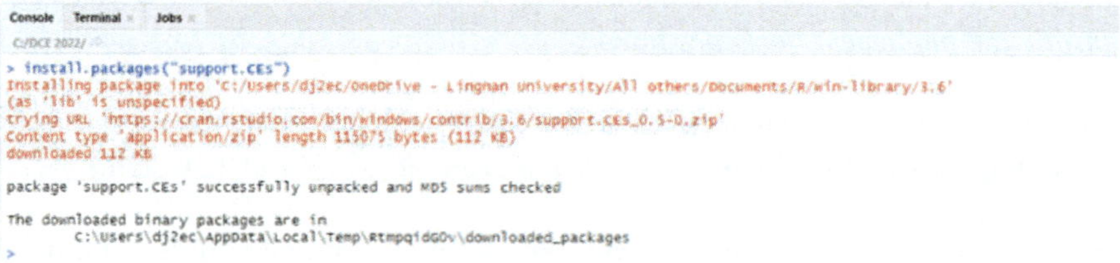

Fig. 4.7 Typing a code to install an R package in RStudio

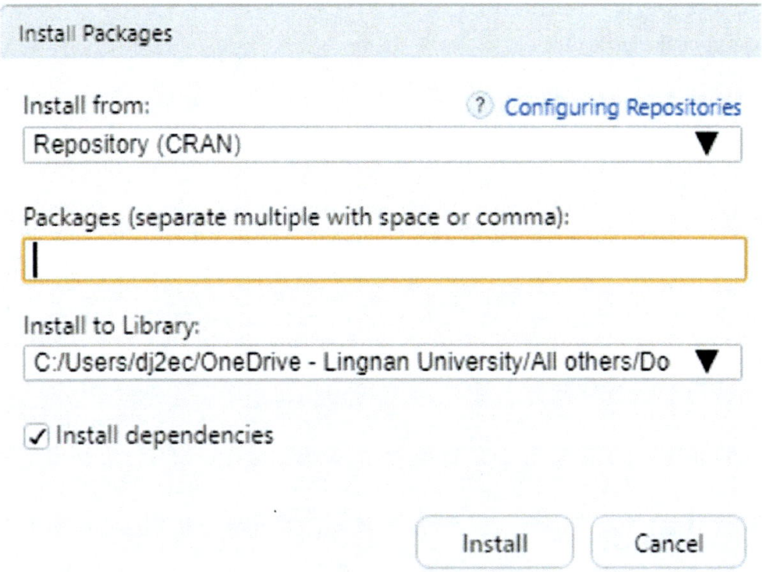

Fig. 4.8 Installing packages from the menu function

4.4 Using R for DCE Experiment

Given the thriving and active R community, we can use many R packages to support DCE studies in various ways, from identifying the attributes to the DCE experimental design, and preparing datasets, to the DCE data analysis and estimation. In this section, we will highlight several R packages that are widely used by academics and briefly illustrate how these could be used to support DCE studies.

4.4.1 Attributes Identification

As mentioned earlier in Chap. 3, identifying relevant attributes to describe the product or service or idea under investigation is the first step in designing a DCE study (Ryan et al., 2001; Soekhai et al., 2019). Thus, attributes identification to construct alternatives for evaluation in DCE studies is vital to the DCE design stage. Various methods are used to identify attributes in a discrete choice experiment. One of the most common methods is conducting a *systematic literature review* to gather information from existing studies (e.g., Hundley et al., 2001; Kim et al., 2020). Another approach involves collecting qualitative data through methods such as *semi-structured interviews* (e.g., Doherty et al., 2021) or *focus group discussions* (e.g., Abiiro et al., 2014; Minnis et al., 2019). Additionally, *experts* can be consulted for their opinions on which attributes are most relevant (e.g., Güney & Giraldo, 2019). In practice,

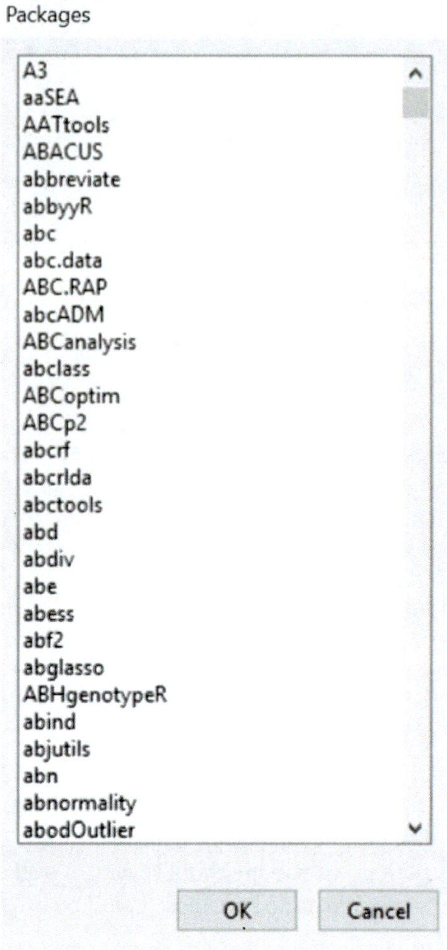

Fig. 4.9 Installing R packages from the menu of R console

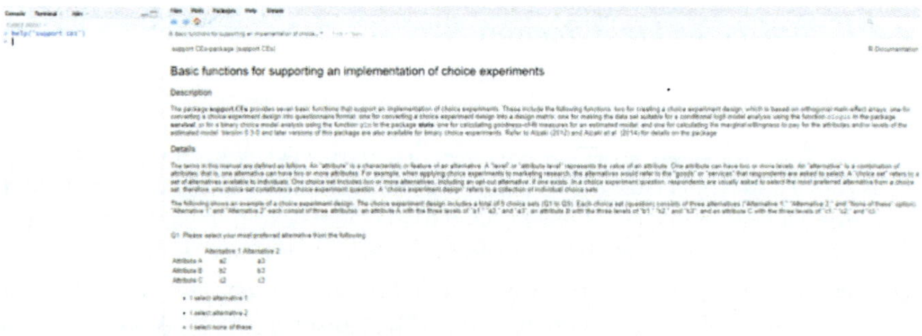

Fig. 4.10 A screenshot of the *help* function

researchers often combine multiple approaches, such as using both literature review and focus group interviews, to identify the most important and influential attributes (e.g., Dubov et al., 2019; Zanolini et al., 2018).

After collecting the necessary secondary and/or primary information, researchers need to conduct a *content analysis* of qualitative data to derive possible attributes for a DCE study. This process often involves carefully reading interview transcripts and identifying major categories or themes that emerge from systematic data analysis. Despite the importance of the attributes identification stage to the overall DCE design, there is scarce guidance and resources to help researchers in attributes selection. We partly offer some insights into this process in this book.

In the R platform, a package called **RQDA** was developed specifically to assist in the *analysis of textual data* (Huang, 2014), such as interview transcripts or media articles. You can read our book on using **RQDA** to conduct a systematic qualitative data analysis (Chandra & Shang, 2019). This package allows researchers to retrieve and code (or assign meanings and categories of) qualitative data in a systematic manner. It is particularly useful when the number of interviews or the amount of qualitative information is large. **RQDA** has a separate window running a graphical user interface (through **RGtk2**[5]) that is easy to manipulate and use. A screenshot of the **RQDA** user interface is shown below in Fig. 4.11.

4.4.2 DCE Experimental Design

After you have identified attributes and attribute levels for a DCE study, the next step is the *experimental design*—generating the alternatives and combining them to form *choice sets*. We identified three R packages that can support researchers in generating experimental designs for DCE studies: "**support.CEs**", "**choiceDes**", and "**idefix**". We will briefly introduce each package below and provide more details with step-by-step instructions in Chap. 7.

When designing a DCE, researchers are usually restricted by the number of choice sets to include in the questionnaire to present to respondents. Parsimony is key; that is, you should not include too many attributes and attribute levels in a DCE to avoid respondent fatigue. "Less is more" is a great philosophy for attributes and attribute levels selection. In various studies we reviewed, researchers who used DCEs have asked respondents to evaluate up to 18 choice sets, a practical threshold for *boredom* (Hanson et al., 2005; Kessels et al., 2011). Therefore, a *full factorial design*, that is the set of all possible choice sets one could present,[6] may become more and more challenging to implement with the increase in the number of attributes and attribute

[5] Due to the update of R platforms, the official published RQDA package has been archived on CRAN. Interested users may need to install the package from the archives or use an older version of R (i.e., R version 3.6.3 or older) for installation.

[6] 3 levels for the price, 2 levels for production, 2 levels for sugar amount, and 3 levels for origin creates a $3 \times 2 \times 2 \times 3$, thus 36 possible profiles.

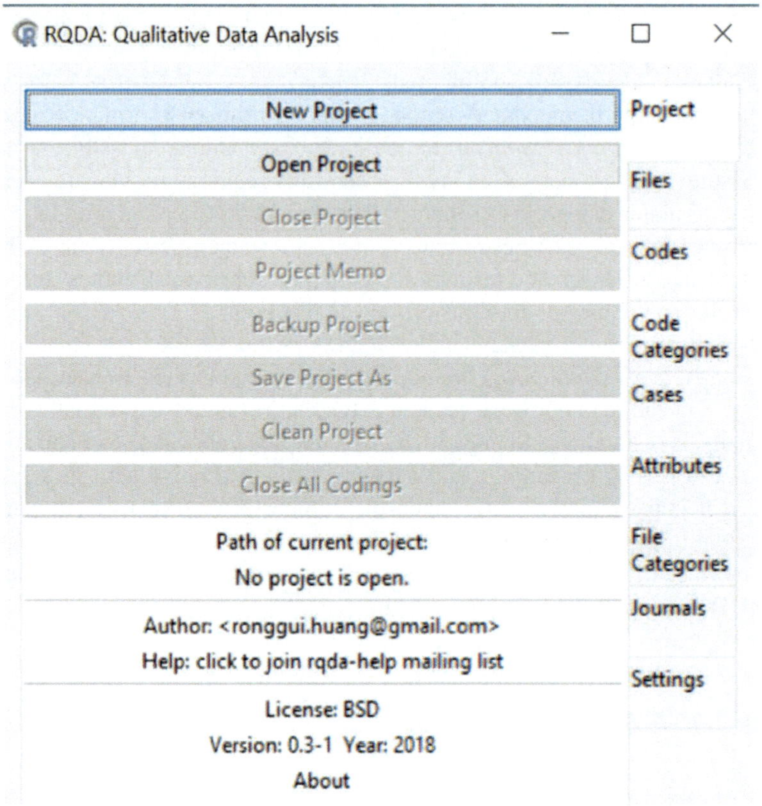

Fig. 4.11 RQDA user interface for textual data analysis

levels used. Thus, researchers need to decide on the number of choice sets they would like to include in a DCE.

support.CEs is an R package that supports the creation of a choice experiment design that is based on orthogonal main-effect arrays (Aizaki, 2012; Aizaki & Nishimura, 2008). Using the orange juice example we provided in Chap. 3 earlier, in which four attributes (i.e., see Table 3.1) were used to describe the product, we can use this package to generate the DCE experimental design. A screenshot of the output produced by the **support.CEs** package is shown in Fig. 4.12.

The generated output shows an orthogonal main-effect design with a total of 12 choice sets, which uses only a fraction of the full factorial design. Each choice set has two alternatives, described by attributes (i.e., "Organic", "Sugar", "Country", and "Price") and attribute levels. Moreover, there is only one block in this design, which means that all respondents will receive the same choice sets.

The package **support.CEs** also supports converting a choice experiment design into a questionnaire format for easier implementation in a particular software (e.g.,

```
Choice sets:
alternative 1 in each choice set
   BLOCK QES ALT       Organic Sugar  Country Price
6      1   1   1        organic   15g    Local     5
1      1   2   1        organic   25g    Local    10
3      1   3   1        organic   25g    Local     5
12     1   4   1 conventional   15g    Local     8
10     1   5   1        organic   15g Imported     8
7      1   6   1 conventional   25g Imported     5
9      1   7   1        organic   15g Imported    10
2      1   8   1 conventional   15g Imported     5
4      1   9   1 conventional   15g    Local    10
8      1  10   1 conventional   25g    Local     8
5      1  11   1 conventional   25g Imported    10
11     1  12   1        organic   25g Imported     8

alternative 2 in each choice set
   BLOCK QES ALT       Organic Sugar  Country Price
6      1   1   2 conventional   25g Imported     8
1      1   2   2 conventional   15g Imported     5
3      1   3   2 conventional   15g Imported     8
12     1   4   2        organic   25g Imported    10
10     1   5   2 conventional   25g    Local    10
7      1   6   2        organic   15g    Local     8
9      1   7   2 conventional   25g    Local     5
2      1   8   2        organic   25g    Local     8
4      1   9   2        organic   25g Imported     5
8      1  10   2        organic   15g Imported    10
5      1  11   2        organic   15g    Local     5
11     1  12   2 conventional   15g    Local    10
```

Fig. 4.12 A DCE design using the package support.CEs

online DCE survey using Qualtrics, Question Pro, or Google Survey) or other conventional approaches (e.g., paper-based survey or telephone survey DCE survey), as shown in Fig. 4.13.

Secondly, **choiceDes** *is* another R package developed by Horne (2018) to generate choice sets for DCE experiments and other types of choice studies such as MaxDiff and other trade-offs. This package depends on the package **AlgDes** and can generate *D-optimal designs* for the *linear model* (McPhedran et al., 2022; Traets et al., 2020). A screenshot of the result of an optimal restricted fractional factorial design using the package is shown below (see Fig. 4.14).

The above screenshot shows an experimental design generated by the package **choiceDes**. In this design, a total of 16 choice sets were split into two blocks (as specified in "vers"), each with eight choice sets. In each choice set, there are two alternatives (also shown as "task").

Last but not least, **idefix** package was developed by Traets et al. (2020) to generate efficient and optimal DCE designs. This package implements D-efficient, Bayesian

Fig. 4.13 A questionnaire transformation by package support.CEs

```
Question 1
        alt.1     alt.2
Organic "organic" "conventional"
Sugar   "15g"     "25g"
Country "Local"   "Imported"
Price   "5"       "8"

Question 2
        alt.1     alt.2
Organic "organic" "conventional"
Sugar   "25g"     "15g"
Country "Local"   "Imported"
Price   "10"      "5"

Question 3
        alt.1     alt.2
Organic "organic" "conventional"
Sugar   "25g"     "15g"
Country "Local"   "Imported"
Price   "5"       "8"

Question 4
        alt.1          alt.2
Organic "conventional" "organic"
Sugar   "15g"          "25g"
Country "Local"        "Imported"
Price   "8"            "10"

Question 5
        alt.1      alt.2
Organic "organic"  "conventional"
Sugar   "15g"      "25g"
Country "Imported" "Local"
Price   "8"        "10"
```

D-efficient, and the IASB approach to generate optimal designs for the *multinomial logit model* (MNL) (McFadden, 1974) and the *mixed logit model* (MIXL) (Hensher & Greene, 2003). The data format of **idefix** package can be easily transformed by using existing estimation packages in R. A screenshot of the design outputs generated using the **idefix** package is shown in Fig. 4.15.

The example above demonstrates a DCE design with three attributes. One attribute has two levels, one has four levels, and the other one has three levels. This design contains a total of 24 profiles, and the screenshot shows each profile described by the three attributes.[7]

[7] Var12 refers to the second level of Attribute 1. Var 21, Var 22, and Var 23 refer to the 2nd, 3rd, and 4th levels of Attribute 2. Var 31 and V 32 refer to the 2nd and 3rd levels of Attribute 3.

```
$levels
   card vers task X1 X2 X3 X4 X5 X6
6     1    1    1  1  1  1  1  2  1
25    1    1    1  1  1  2  2  1  3
1     2    1    2  2  2  1  1  1  1
8     2    1    2  2  2  1  2  2  1
17    3    1    3  2  1  2  1  2  2
19    3    1    3  1  1  1  2  2  2
15    4    1    4  1  2  2  2  1  2
13    4    1    4  1  1  1  2  1  2
29    5    1    5  1  1  2  1  2  3
27    5    1    5  2  1  1  1  2  3
21    6    1    6  2  2  1  1  1  3
18    6    1    6  1  2  2  1  2  2
26    7    1    7  2  2  2  2  1  3
23    7    1    7  2  2  2  1  1  3
9     8    1    8  2  1  2  2  2  1
5     8    1    8  1  2  1  2  1  1
12    9    2    1  1  2  1  1  1  2
2     9    2    1  1  1  2  1  1  1
14   10    2    2  2  1  2  2  1  2
28   10    2    2  1  2  1  1  2  3
31   11    2    3  1  2  1  2  2  3
3    11    2    3  1  2  2  1  1  1
22   12    2    4  1  1  2  1  1  3
11   12    2    4  2  1  1  1  1  2
24   13    2    5  2  1  1  2  1  3
20   13    2    5  2  2  2  2  2  2
4    14    2    6  2  1  1  2  1  1
32   14    2    6  2  2  2  2  2  3
7    15    2    7  2  1  2  1  2  1
30   15    2    7  1  1  1  2  2  3
10   16    2    8  1  2  2  2  2  1
16   16    2    8  2  2  1  1  2  2
```

Fig. 4.14 A DCE design using the package choiceDes

4.4.3 DCE Data Analysis and Estimation

After DCE data collection, researchers need to transform the collected data into the *correct format* to use estimation packages in R to analyze the data. Some DCE experimental design packages could support the transformation of datasets into suitable formats. For example, the package **support.CEs** has a function to make a data set suitable for a conditional logit model analysis in the R package called **survival** (Therneau & Lumley, 2015) or for a binary choice model analysis in the core R package **stats**. Similarly, **idefix** also offers ***Datatrans()*** function to support the data

Fig. 4.15 A DCE design using package idefix

```
Console   Terminal   Jobs
C:/DCE 2022/
> library(idefix)
> at.lvls <- c(2,4,3)
> c.type <- c("D","E","E")
> Profiles(lvls = at.lvls, coding = c.type)
   Var12 Var21 Var22 Var23 Var31 Var32
1      0     1     0     0     1     0
2      1     1     0     0     1     0
3      0     0     1     0     1     0
4      1     0     1     0     1     0
5      0     0     0     1     1     0
6      1     0     0     1     1     0
7      0    -1    -1    -1     1     0
8      1    -1    -1    -1     1     0
9      0     1     0     0     0     1
10     1     1     0     0     0     1
11     0     0     1     0     0     1
12     1     0     1     0     0     1
13     0     0     0     1     0     1
14     1     0     0     1     0     1
15     0    -1    -1    -1     0     1
16     1    -1    -1    -1     0     1
17     0     1     0     0    -1    -1
18     1     1     0     0    -1    -1
19     0     0     1     0    -1    -1
20     1     0     1     0    -1    -1
21     0     0     0     1    -1    -1
22     1     0     0     1    -1    -1
23     0    -1    -1    -1    -1    -1
24     1    -1    -1    -1    -1    -1
```

transformation. In particular, ***Datatrans()*** allows researchers to specify the format needed for the datasets according to the package that will be used for estimation.

There is a wide range of packages available in R that support discrete choice data estimation. Besides the ones mentioned above, there are also **bayesm** (Rossi, 2019), **Rchoice** (Sarrias, 2016), **ChoiceModelR** (Sermas, 2012), **RSGHB** (Dumont & Keller, 2019), and **mlogit** (Croissant, 2020).

To illustrate it further (see Fig. 4.16), we provide a screenshot of the DCE data analysis outputs generated using a package called **survival**. We will further discuss the DCE data analysis using different R packages in Chap. 9 later.

Overall, many R packages are available to support DCE experimental design and data analyses in various ways. In the next chapters, we will discuss in more detail how to utilize different R packages to assist the design, implementation, and analysis of a DCE experiment using an example.

```
Call:
clogit(RES ~ ASC + Limited + Task + Clients + Salary + Limited.1 +
    Limited.2 + strata(STR), data = dataset2)

               coef exp(coef) se(coef)      z        p
ASC         2.99910  20.06756  0.25609 11.711  < 2e-16
Limited    -0.34937   0.70513  0.09208 -3.794 0.000148
Task       -0.17086   0.84294  0.11474 -1.489 0.136467
Clients    -0.33427   0.71586  0.10757 -3.107 0.001887
Salary      0.36964   1.44721  0.01601 23.081  < 2e-16
Limited.1  -0.61610   0.54004  0.11695 -5.268 1.38e-07
Limited.2  -0.29278   0.74618  0.11904 -2.460 0.013909
```

Fig. 4.16 DCE data analysis using the package *survival*

References

Abiiro, G. A., Leppert, G., Mbera, G. B., Robyn, P. J., & De Allegri, M. (2014). Developing attributes and attribute-levels for a discrete choice experiment on micro health insurance in rural Malawi. *BMC Health Services Research, 14*(1), 1–15.

Aizaki, H. (2012). Basic functions for supporting an implementation of choice experiments in R. *Journal of Statistical Software, 50*, 1–24.

Aizaki, H., & Nishimura, K. (2008). Design and analysis of choice experiments using R: A brief introduction. *Agricultural Information Research, 17*(2), 86–94.

Chandra, Y., & Shang, L. (2019). *Qualitative research using R: A systematic approach.* Springer.

Croissant, Y. (2020). Estimation of random utility models in R: The mlogit package. *Journal of Statistical Software, 95*, 1–41.

Doherty, E., Mellett, S., Norton, D., McDermott, T. K., O'Hora, D., & Ryan, M. (2021). A discrete choice experiment exploring farmer preferences for insurance against extreme weather events. *Journal of Environmental Management, 290*, 112607.

Dubov, A., Ogunbajo, A., Altice, F. L., & Fraenkel, L. (2019). Optimizing access to PrEP based on MSM preferences: Results of a discrete choice experiment. *AIDS Care, 31*(5), 545–553.

Dumont, J., & Keller, J. (2019). RSGHB: Functions for hierarchical Bayesian estimation: A flexible approach. R package version 1.2.2. https://CRAN.R-project.org/package= RSGHB

Güney, O. I., & Giraldo, L. (2019). Consumers' attitudes and willingness to pay for organic eggs: A discrete choice experiment study in Turkey. *British Food Journal.*

Hanson, K., McPake, B., Nakamba, P., & Archard, L. (2005). Preferences for hospital quality in Zambia: Results from a discrete choice experiment. *Health Economics, 14*(7), 687–701.

Hensher, D. A., & Greene, W. H. (2003). The mixed logit model: The state of practice. *Transportation, 30*(2), 133–176.

Huang, R. (2014). *RQDA: R-based qualitative data analysis.* https://rqda.r-forge.r-project.org/

Hundley, V., Ryan, M., & Graham, W. (2001). Assessing women's preferences for intrapartum care. *Birth, 28*(4), 254–263.

Kessels, R., Jones, B., & Goos, P. (2011). Bayesian optimal designs for discrete choice experiments with partial profiles. *Journal of Choice Modelling, 4*(3), 52–74.

Kim, J., Lee, H., & Lee, J. (2020). Smartphone preferences and brand loyalty: A discrete choice model reflecting the reference point and peer effect. *Journal of Retailing and Consumer Services, 52*, 101907.

McFadden, D. (1974). The measurement of urban travel demand. *Journal of Public Economics, 3*(4), 303–328.

McPhedran, R., Gold, N., Bemand, C., Weston, D., Rosen, R., Scott, R., ... & Toombs, B. (2022). Location, location, location: A discrete choice experiment to inform COVID-19 vaccination programme delivery in the UK. *BMC public health, 22*(1), 1–12.

Minnis, A. M., Browne, E. N., Boeri, M., Agot, K., Van Der Straten, A., Ahmed, K., ... & Mansfield, C. (2019). Young women's stated preferences for biomedical HIV prevention: Results of a discrete choice experiment in Kenya and South Africa. *Journal of Acquired Immune Deficiency Syndromes (1999)*, *80*(4), 394.

Rossi, P. (2019). *bayesm: Bayesian inference for marketing/micro-econometrics.* R package version 3.1–4. https://CRAN.R-project.org/package=bayesm

RStudio Team. (2021). *RStudio: Integrated development for R.* http://www.rstudio.com/

Ryan, M., Bate, A., Eastmond, C. J., & Ludbrook, A. (2001). Use of discrete choice experiments to elicit preferences. *BMJ Quality & Safety, 10*(suppl 1), i55–i60.

Sarrias, M. (2016). Discrete choice models with random parameters in R: The Rchoice package. *Journal of Statistical Software, 74*, 1–31.

Sermas, R. (2012). *ChoiceModelR: Choice modeling in R.* R package version 1.2. https: // CRAN.R-project.org/package=ChoiceModelR

Soekhai, V., de Bekker-Grob, E. W., Ellis, A. R., & Vass, C. M. (2019). Discrete choice experiments in health economics: Past, present and future. *PharmacoEconomics, 37*(2), 201–226.

Therneau, T. M., & Lumley, T. (2015). Package 'survival.' *R Top Doc, 128*(10), 28–33.

Traets, F., Sanchez, D. G., & Vandebroek, M. (2020). Generating optimal designs for discrete choice experiments in R: The Idefix package. *Journal of Statistical Software, 96*, 1–41.

Zanolini, A., Sikombe, K., Sikazwe, I., Eshun-Wilson, I., Somwe, P., Bolton Moore, C., ... & Geng, E. H. (2018). Understanding preferences for HIV care and treatment in Zambia: evidence from a discrete choice experiment among patients who have been lost to follow-up. *PLoS Medicine, 15*(8), e1002636.

Chapter 5
Framing the Research Question

DCE experiments should ideally be grounded on a research question or puzzles. A puzzle may refer to an unresolved or debated question in a specific discipline or topic. *Accurately framing the research question* is crucial in determining the subsequent decisions in the research design, such as the number and types of attributes and attribute levels to be included and the methods used to elicit decision-making from respondents. This chapter discusses how to define a specific research question and ground it in existing theories or literature. Some DCE studies are preceded by hypotheses (following "hypothetico deductive" science where theory or literature on the topic is available to form predictions), while many other DCE studies do not explicitly use hypotheses because researchers use it as a form of quantitative exploration (Wennberg & Anderson, 2020). We show how research questions can be developed using easy-to-follow examples to suit the DCE experiment.

5.1 Types of Research Questions That Are Suitable for DCEs

DCE has become increasingly popular because it offers opportunities for researchers to *address a wide range of research questions*, some of which cannot be otherwise answered adequately using other approaches. Generally speaking, DCEs can be used for both *exploratory* and *explanatory* studies, depending on the research question and the study context. For exploratory studies that generate new insights or develop new theories, DCEs can be used to identify which attributes or levels are important in influencing choices without prior assumptions or hypotheses. DCEs can also explore how different groups of people make choices and generate hypotheses for further research. In this context, DCEs can be useful for theory-building and exploratory research. For explanatory studies that aim at theory testing or theory elaboration, DCEs can be used to test the impact of specific attributes or levels on choices and to explain the underlying mechanisms that drive choices.

L. Shang and Y. Chandra, *Discrete Choice Experiments Using R*,
https://doi.org/10.1007/978-981-99-4562-7_5

Moreover, DCEs are particularly useful for answering what and which questions. By presenting respondents with a series of scenarios and asking them to choose between different scenarios, DCEs can reveal the relative importance of different attributes and levels that influence choices. However, DCEs are less suited to answering why questions, as they do not provide direct insights into the underlying reasons for respondents' choices. That said, DCEs can be used in conjunction with other methods, such as interviews or surveys, to explore the reasons behind respondents' choices.

Based on our literature review of existing DCE studies, we summarized five major research questions and objectives that could be addressed using a DCE methodology. We present each type respectively below.

5.1.1 Preference for Different Types of Services/Products

The first type of research question and objective concerns the evaluation of respondents' *Preference for different products or services* and how different attributes could shape individuals' decision-making. This type of research question typically *juxtaposes two or more alternatives* to elicit which one is more preferred by respondents. For instance, Ewing and Sarigöllü (2000) designed a DCE to examine user preference for clean-fuel vehicles versus conventional vehicles. In their study, respondents were asked to choose whether they prefer to buy a clean-fuel vehicle (i.e., low-emission vehicles vs. zero-emission vehicles) or a conventional vehicle with different conditions.

To answer research question of this type, researchers usually present respondents with choice sets of two or more different types of products or services and are asked to indicate their preference. In this type of study, researchers can examine how the decisions of individuals would change from one product or service to another by varying the values in different attributes (e.g., 5 g sugar vs. 10 g sugar in a yogurt product). Some more examples of research questions and objectives that belong to this type include:

- Blaauw et al. (2010): this evaluates the relative effectiveness of different policies in attracting nurses to rural areas in Kenya, South Africa, and Thailand (research question: *Which policy is most preferred?*)
- Fu et al. (2020): this elicits the Preference of potential travelers for hotels and peer-to-peer accommodation sharing on online booking platforms (research question: *Do users prefer hotels or peer-to-peer accommodation?*)
- Bjørnåvold et al. (2020): this assesses whether European policymakers prefer to invest in technologies that resemble the incumbent or in novel and more disruptive technologies (research question: *Whether the disruptive or the incumbent technology is more preferred by policymakers)*

- Hjelmgren and Anell (2007): this examines whether individuals who were given a choice preferred individual family physicians or a primary care team consisting of physicians and nurses (research question: *Whether individual family physicians or a primary care team is more preferred by individuals?*).

5.1.2 Attribute Evaluation by Decision-Makers

The second type of research question and objective that can be addressed using DCEs focuses on *quantifying the trade-offs* that decision-makers are prepared to make between different aspects of a product, service, or idea to elicit *which attributes are most valued* by participants. For instance, Mangham and Hanson (2008) conducted a DCE study to understand the employment preference of Malawian public sector registered nurses. Their study examined how nurses make trade-offs among six job attributes, including monthly pay, availability of government housing, opportunities to upgrade their qualifications, typical workload, availability of resources, and place of work.

This type of DCE study investigates how individuals make trade-offs in decision-making. This provides insights into the relative importance and value individuals assign to different attributes of a certain product or service, thus allowing practitioners and policymakers to optimize resource allocation or policy design. Other examples include the following:

- Albaladejo-Pina and Díaz-Delfa (2009): this evaluates tourists' Preference for different attributes (e.g., size, type of building, quality of equipment) of rural houses (research question: *What trade-offs do tourists make when choosing a rural house?*)
- Cleland et al. (2016): this investigates the strength of UK foundation doctors' and trainees' Preference for various training post characteristics in terms of monetary value (research question: *What are the most important characteristics of a training post to doctors and trainees?)*
- Turner et al. (2007): this estimates the relative importance of continuity of care compared to other factors during a primary care consultation from the perspective of patients (research question: *Which are the most important aspects of a primary care consultation to patients?)*
- Giles et al. (2016): this studies the relative Preference of UK adults for different attributes of financial incentives for healthy behaviors (research question: *Which are the most valued attributes of financial incentives by UK adults?*).

5.1.3 *Monetary Valuation of Attributes*

The third type of DCE research question and objective pertains to *evaluating the decision-maker's preference in monetary terms*. Besides estimating each attribute's relative importance and strength, DCEs can also be used to calculate the *willingness to pay* (WTP) for a unit change in each attribute—that is, the maximum price an individual is willing to pay for positive changes in certain attributes. In this type of study, a monetary attribute, such as prices, salaries, and financial incentives, is normally included in the DCE design to determine individuals' WTP values for different attributes and attribute levels (Alpizar et al., 2003).

For instance, Raffaelli and colleagues (2022) investigated tourists' preferences and WTP values for decarbonization strategies applied to transportation and hotel accommodation. Their study demonstrates that tourists attach low to zero WTP values to the use of electric trains and the possibility of offsetting the carbon emissions associated with their hotel stays, suggesting that tourists are not willing to bear the costs of sustainable behavior. Other examples are provided below:

- Masiero et al. (2015): this determines guests' willingness to pay for different hotel room attributes within a single hotel property (research question: *How much are guests willing to pay for different hotel facilities?*)
- Nieboer et al. (2010): this examines the population's willingness to pay for different attributes of long-term care services (research question: *How much are old adults willing to pay for different conditions of long-term care?*)
- Meenakshi et al. (2012): this estimates the willingness to pay for biofortified orange maize in rural Zambia with different marketing and campaign approaches (research question: *How do nutrition campaigns change residents' willingness to pay for orange maize?*)
- Wang et al. (2018): this investigates Chinese consumers' willingness to pay for pork with certified labels (research question: *How much will consumers pay for different food safety certification labels?*).

5.1.4 *Estimating Acceptance for Undesirable Changes*

Next, DCEs can also be utilized to estimate individuals' acceptance of a new initiative deemed undesirable or containing *potential risks or costs*, such as new health treatment, vaccination, new policy regulation, and the development of harmful infrastructure. For this type of study, researchers normally present respondents with "scenarios" that illustrate both benefits and risks of a new initiative and ask respondents if they are willing to change their status quo and accept the new initiative under different conditions. By changing the values of different attributes, researchers could obtain insights into what conditions could motivate individuals to accept new changes or uptake a new program.

For example, Dong and colleagues (2020) presented a DCE study that examined citizens' acceptance of the COVID-19 vaccination program in China. The authors argued that public acceptance is considered a vital factor for successfully adopting any vaccination program. The findings of their DCE study indicated that the *effectiveness of the vaccine* is the most influential attribute that could increase public acceptance and improve the public motivation to participate in COVID-19 vaccination. Insights from these DCE studies could be used to *forecast demand* (e.g., forecasting uptake of vaccines) and assist in planning appropriate levels of provision and implementation. Other examples of research include:

- McNamara et al. (2013): this investigates factors that influence older people in the decision to relinquish their driver's license (research question: *What factors will motivate older people to give up their driver's license?*)
- Jonker et al. (2020): this estimates the future uptake of a smartphone-based contact tracing app in the Dutch population (research question: *Which attributes of the contact tracing app will motivate citizens' adoption?*)
- Manipis et al. (2021): this explores the acceptability of different COVID-19 control measures (research question: *Which factors will influence public acceptance toward disease control measures?*)
- Degeling et al. (2020): this elicits the preference and choice trends for attributes associated with technologically enhanced communicable disease surveillance (research question: *Which factors will influence public acceptance towards technologically enhanced surveillance?).*

5.1.5 Unpacking Preference Heterogeneity of Differing Groups

Another type of DCE research question addresses *preference heterogeneity among different populations* by studying individual decisions and preference disparities. In this type of study, data collected from a DCE are analyzed by *sub-groups*. Thus, it is possible to estimate the *extent to which individual characteristics influence marginal valuations* (e.g., Scarpa et al., 2005). Such insights could be useful for policymakers or practitioners to take target measures to tailor the provision of products or services for different groups of populations.

For instance, the DCE study by Mentzakis and colleagues (2011) explored the preference heterogeneity in informal care. They found that *monetary compensation* is more important for young people but insignificant for older adults. Similarly, Jiang et al.'s (2020) DCE study also focused on identifying preference heterogeneity among residents for primary health services in China. Based on their DCE findings, the authors classified residents into four different groups, each with a particular preference for healthcare services. Other examples are highlighted below:

- Bechtold & Abdulai (2014): this examines the preference for functional dairy product attributes and estimates the preference heterogeneity (research question: *What are the differences in consumers' preference for functional dairy products?*)
- Song et al. (2015): this studies which job attributes affect Chinese primary care providers' choice of job and whether there are any differences in the job preference between doctors and nurses (research question: *How do job preferences of doctors and nurses differ in China?*)
- Buckell et al. (2019): this estimates the preference and demand for cigarettes and e-cigarettes among adult smokers and recent quitters (research question: *What are the differences in the preference for cigarette flavors between adult smokers and recent quitters?*)
- Kjær and Gyrd-Hansen (2008): this elicits patients' preference for cardiac rehabilitation activities and estimates the preference heterogeneity among patients by taking age and smoking status into account (research question: *How do Preference for cardiac rehabilitation activities differ among individuals with different ages and different smoking status?*).

5.2 Defining DCE Research Questions and Objectives

Defining a clear and feasible research question that could be addressed by DCEs is the first step in DCE studies. As mentioned earlier, the design of a DCE study largely depends on the research questions and objectives defined by the researchers (Mangham et al., 2009). To better define research questions suitable using DCE methodology, we propose a framework for researchers when framing their research questions and objectives (see Fig. 5.1).

Importantly, to clearly define research questions and objectives, researchers must first determine the product, service, or policy of interest for examination in light of important debates in a discipline or practice. For example, researchers might be interested in assessing citizens' preference for a *new taxation scheme*. Then the taxation scheme will be the focus of examination in the DCE study. After deciding on the focus of examination, the next question concerns the *target populations* of the DCE study—to whom the DCE questionnaires will be distributed and to which groups the results will be generalized. Researchers should also consider whether *subgroup differences* are a focus of the research or not. For instance, some researchers may be interested in comparing how young and older adults value job opportunities. Hence, in this example, age should be considered in the experimental design as an influential factor.

The final question that researchers should consider pertains to the *decision to be made by respondents*. Researchers need to picture the decision-making scenarios that individuals may encounter in the *real world* and consider the *choice situation* they want to present to respondents in a DCE study. For example, suppose researchers

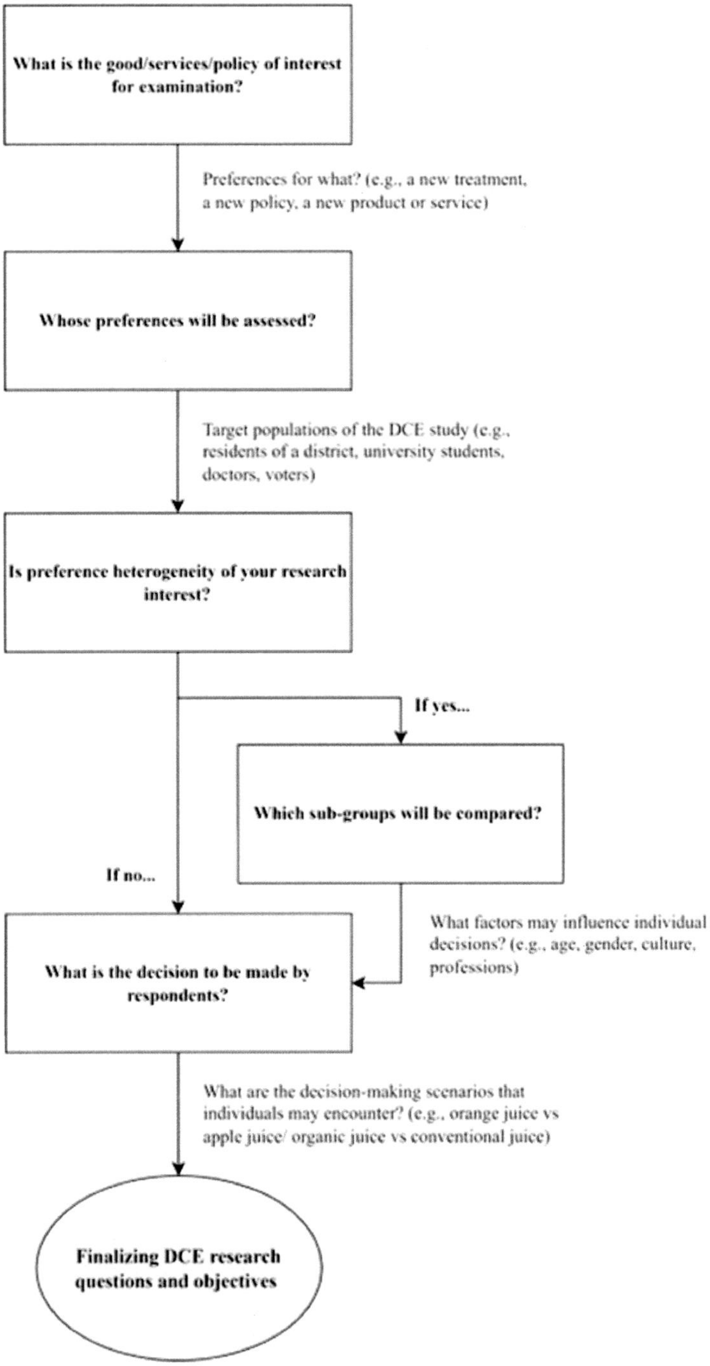

Fig. 5.1 A framework for defining DCE questions and objectives

are interested in studying the preference for organic orange juice. In that case, they need to consider whether individuals will make a decision among different orange juice products or between orange juice and apple juice or juices versus other drinks (e.g., water, soda, liquor). Thus the scope of decisions to examine is important.

To better demonstrate the process of designing and implementing a DCE study using a step-by-step "hand-holding" approach, we use our sample research project, *Examining Consumer Preference for Alternative Meat*, as an illustration for this book. In this sample project, we are interested in examining consumer preference for alternative meat and plant-based food products made entirely from plants or vegan ingredients. Our *target populations* are general customers who are aged 18 or above. Besides exploring which attributes of alternative meat products are important to customers in general, we were also curious about individual differences in decision-making regarding alternative meat. Finally, the *choice situation* we decide is to let respondents *choose between two alternative meat products* instead of between alternative meat products and conventional meat products. This approach will allow us to better understand important considerations and trade-offs of individuals in choosing alternative meat products.

Consequently, we proposed the following research question in this DCE research project: *What are customers' preferences for different alternative meat products?*

The key research objectives of this sample project are:

- To investigate customers' Preference for purchasing alternative meat products;
- To identify which attributes of alternative meat products affect customers' purchasing decisions; and
- To examine whether individual characteristics, such as gender, educational level, and environmental awareness, affect customers' Preference for alternative meat products or not.

After finalizing the research questions and objectives, we could follow the different stages of designing a DCE study as outlined in Chap. 3 (see Fig. 3.1) by identifying attributes to be included in the DCE study. In the next chapter, we will further discuss different approaches for identifying DCE attributes and important principles in selecting suitable attributes.

References

Albaladejo-Pina, I. P., & Díaz-Delfa, M. T. (2009). Tourist preference for rural house stays: Evidence from discrete choice modelling in Spain. *Tourism Management, 30*(6), 805–811.

Alpizar, F., Carlsson, F., & Martinsson, P. (2003). Using choice experiments for non-market valuation. *Economic Issues Journal Articles, 8*(1), 83–110.

Bechtold, K. B., & Abdulai, A. (2014). Combining attitudinal statements with choice experiments to analyze preference heterogeneity for functional dairy products. *Food Policy, 47*, 97–106.

Bjørnåvold, A., Lizin, S., Van Dael, M., Arnold, F., & Van Passel, S. (2020). Eliciting policy-makers' Preference for technologies to decarbonise transport: A discrete choice experiment. *Environmental Innovation and Societal Transitions, 35*, 21–34.

Blaauw, D., Erasmus, E., Pagaiya, N., Tangcharoensathein, V., Mullei, K., Mudhune, S., ... & Lagarde, M. (2010). Policy interventions that attract nurses to rural areas: a multicountry discrete choice experiment. *Bulletin of the World Health Organization, 88*(5), 350–356.

Buckell, J., Marti, J., & Sindelar, J. L. (2019). Should flavours be banned in cigarettes and e-cigarettes? Evidence on adult smokers and recent quitters from a discrete choice experiment. *Tobacco Control, 28*(2), 168–175.

Cleland, J., Johnston, P., Watson, V., Krucien, N., & Skåtun, D. (2016). What do UK doctors in training value in a post? A discrete choice experiment. *Medical Education, 50*(2), 189–202.

Degeling, C., Chen, G., Gilbert, G. L., Brookes, V., Thai, T., Wilson, A., & Johnson, J. (2020). Changes in public Preference for technologically enhanced surveillance following the COVID-19 pandemic: A discrete choice experiment. *British Medical Journal Open, 10*(11), e041592.

Dong, D., Xu, R. H., Wong, E. L. Y., Hung, C. T., Feng, D., Feng, Z., ... & Wong, S. Y. S. (2020). Public preference for COVID-19 vaccines in China: A discrete choice experiment. *Health Expectations, 23*(6), 1543–1578.

Ewing, G., & Sarigöllü, E. (2000). Assessing consumer preference for clean-fuel vehicles: A discrete choice experiment. *Journal of Public Policy & Marketing, 19*(1), 106–118.

Fu, S., Cheng, X., Bao, Y., Bilgihan, A., & Okumus, F. (2020). Staying in a hotel or peer-to-peer accommodation sharing? A discrete choice experiment with online reviews and discount strategies. *Internet Research: Electronic Networking Applications and Policy, 31*(2), 654–676.

Giles, E. L., Becker, F., Ternent, L., Sniehotta, F. F., McColl, E., & Adams, J. (2016). Acceptability of financial incentives for health behaviours: A discrete choice experiment. *PLoS ONE, 11*(6), e0157403.

Hjelmgren, J., & Anell, A. (2007). Population preference and choice of primary care models: A discrete choice experiment in Sweden. *Health Policy, 83*(2–3), 314–322.

Jiang, M. Z., Fu, Q., Xiong, J. Y., Li, X. L., Jia, E. P., Peng, Y. Y., & Shen, X. (2020). Preference heterogeneity of health care utilization of community residents in China: A stated preference discrete choice experiment. *BMC Health Services Research, 20*(1), 1–11.

Jonker, M., de Bekker-Grob, E., Veldwijk, J., Goossens, L., Bour, S., & Rutten-Van Mölken, M. (2020). COVID-19 contact tracing apps: Predicted uptake in the Netherlands based on a discrete choice experiment. *JMIR mHealth and uHealth, 8*(10), e20741.

Kjaer, T., & Gyrd-Hansen, D. (2008). Preference heterogeneity and choice of cardiac rehabilitation program: Results from a discrete choice experiment. *Health Policy, 85*(1), 124–132.

Mangham, L. J., & Hanson, K. (2008). Employment preference of public sector nurses in Malawi: Results from a discrete choice experiment. *Tropical Medicine & International Health, 13*(12), 1433–1441.

Mangham, L. J., Hanson, K., & McPake, B. (2009). How to do (or not to do)... Designing a discrete choice experiment for application in a low-income country. *Health Policy and Planning, 24*(2), 151–158.

Manipis, K., Street, D., Cronin, P., Viney, R., & Goodall, S. (2021). Exploring the trade-off between economic and health outcomes during a pandemic: A discrete choice experiment of lockdown policies in Australia. *The Patient-Patient-Centered Outcomes Research, 14*(3), 359–371.

Masiero, L., Heo, C. Y., & Pan, B. (2015). Determining guests' willingness to pay for hotel room attributes with a discrete choice model. *International Journal of Hospitality Management, 49*, 117–124.

McNamara, A., Chen, G., George, S., Walker, R., & Ratcliffe, J. (2013). What factors influence older people in the decision to relinquish their driver's licence? A discrete choice experiment. *Accident Analysis & Prevention, 55*, 178–184.

Meenakshi, J. V., Banerji, A., Manyong, V., Tomlins, K., Mittal, N., & Hamukwala, P. (2012). Using a discrete choice experiment to elicit the demand for a nutritious food: Willingness-to-pay for orange maize in rural Zambia. *Journal of Health Economics, 31*(1), 62–71.

Mentzakis, E., Ryan, M., & McNamee, P. (2011). Using discrete choice experiments to value informal care tasks: Exploring preference heterogeneity. *Health Economics, 20*(8), 930–944.

Nieboer, A. P., Koolman, X., & Stolk, E. A. (2010). Preference for long-term care services: Willingness to pay estimates derived from a discrete choice experiment. *Social Science & Medicine, 70*(9), 1317–1325.

Raffaelli, R., Franch, M., Menapace, L., & Cerroni, S. (2022). Are tourists willing to pay for decarbonizing tourism? Two applications of indirect questioning in discrete choice experiments. *Journal of Environmental Planning and Management, 65*(7), 1240–1260.

Scarpa, R., Philippidis, G., & Spalatro, F. (2005). Product-country images and preference heterogeneity for Mediterranean food products: a discrete choice framework. *Agribusiness: An International Journal, 21*(3), 329–349.

Song, K., Scott, A., Sivey, P., & Meng, Q. (2015). Improving Chinese primary care providers' recruitment and retention: A discrete choice experiment. *Health Policy and Planning, 30*(1), 68–77.

Turner, D., Tarrant, C., Windridge, K., Bryan, S., Boulton, M., Freeman, G., & Baker, R. (2007). Do patients value continuity of care in general practice? An investigation using stated preference discrete choice experiments. *Journal of Health Services Research & Policy, 12*(3), 132–137.

Wang, J., Ge, J., & Ma, Y. (2018). Urban Chinese consumers' willingness to pay for pork with certified labels: A discrete choice experiment. *Sustainability, 10*(3), 603.

Wennberg, K., & Anderson, B. S. (2020). Enhancing the exploration and communication of quantitative entrepreneurship research. *Journal of Business Venturing, 35*(3), 105938.

Chapter 6
Identifying DCE Attributes and Levels

The next stage of DCE research design involves *selecting attributes* (e.g., salary, work hours) in an effort to address the research question and *assigning levels* for each attribute (e.g., 25% higher than average salary vs. average salary vs. 10% lower than average salary; 10-h work or 12-h work). The validity of a DCE study depends on the researcher's ability to properly specify and theoretically justify the relevant attributes and attribute levels. This chapter discusses the principles, approaches, and techniques to select an appropriate number of attributes and attribute levels using an actual research project that we did. We provide step-by-step instructions on how to establish attributes and levels in a DCE study.

6.1 Establishing DCE Attributes

Discrete choice experiments (DCEs) assume that a good or service can be described by its attributes, and individuals' valuation of choices depends upon the levels of these attributes (Abiiro et al., 2014). Inappropriate selection of attributes or misspecification of attributes and levels could lead to unintended implications for the DCE design and implementation, producing inaccurate or confusing results (Mangham et al., 2009). Based on our observation and experience, many scholars using DCE studies did not provide sufficient information and theoretical justification on how they establish the attributes and levels (Coast & Horrocks, 2007). Hence, there have been calls for more rigorous, systematic, and transparent approaches to developing DCE attributes and levels. Our book is one of the first attempts to address this gap.

In the following sections, we will discuss several important questions regarding the development and selection of attributes in a DCE study, followed by a step-by-step illustration from a sample project of ours:

- What are good DCE attributes?
- What are different approaches to identifying attributes?
- How should we assign levels to attributes?

L. Shang and Y. Chandra, *Discrete Choice Experiments Using R*,
https://doi.org/10.1007/978-981-99-4562-7_6

6.1.1 Characteristics of Good Attributes in DCE Studies

Poorly selected attributes will lead to invalid results in DCE studies (Pérez-Troncoso, 2020). Selecting and defining the appropriate attributes require researchers to have a good understanding of the *target respondents' experiences* and the *decision-making scenarios* they encounter in real life (Kjaer, 2005). We highlight five important criteria that help researchers in selecting attributes and attribute levels:

1. *Relevance*: the chosen attributes should be relevant to the product or service under investigation and to the context of the DCE study;
2. *Significance*: the chosen attributes should be important and meaningful for individuals in coming to a decision;
3. *Clarity*: the chosen attributes should be defined in a specific way to avoid attribute ambiguity;
4. *Completeness*: the chosen attributes should be inclusive and cover all important aspects of the product or service under investigation;
5. *Measurability*: the chosen attributes should be measurable either qualitatively or quantitatively.

First, the selected attributes should be relevant to the research question and to the good or service under examination. In most cases, an attribute is considered relevant if neglecting it would result in a different set of conclusions (Kjaer, 2005). However, in some cases, an attribute could be relevant to the research question but might be *demand-irrelevant*. For example, for consumers with low or zero environmental awareness, an attribute that concerns the level of carbon emission of electronic vehicles would be viewed as demand-irrelevant. However, for consumers that care about environmental protection, an environment-related attribute will be very relevant to individuals' decision-making (Bennett & Blamey, 2001). Therefore, to select appropriate attributes, researchers must clearly *specify the target populations* for their DCE studies and develop a good understanding of target respondents' perspectives and experiences regarding the product or service or idea under investigation.

Second, the selected attributes must be considered important by the target population, whose preference will be elicited in DCEs and with consideration of their needs in the local context (Bennett & Blamey, 2001). In particular, an attribute should be meaningful to participants and is able to influence their decision-making process. Certainly, it is not possible and feasible for a DCE study to include every attribute important to each respondent but it is critical to capture attributes that are *salient* to the majority of the target population (Kløjgaard et al., 2012). Notably, in some cases, an attribute may be considered important by researchers or of interest to policy-makers but not necessarily important from the respondents' perspective. A rigorous qualitative approach could be used to assist in deriving meaningful and important attributes for target respondents.

Third, the selected attributes should be defined and described in a specific way to avoid confusion and attribute ambiguity. Unclear attributes and levels could lead to different interpretations of attribute meanings and confounding attribute effects

(Coast et al., 2012; Ryan et al., 2008). Moreover, the definitions of attributes should be appropriate for the local contexts and research setting. For instance, "*compensatory hours*" is a popular term used by employees and employees in Hong Kong when referring to obtaining paid time off to balance out their extra hours of work as an alternative to overtime payment. Whether or not employers allow employees to claim compensatory hours is an important consideration for individuals when making a job decision. Researchers interested in examining job preferences of Hong Kong populations should use *context-specific and familiar expressions* (e.g., compensatory hours) to target respondents to communicate the meanings of attributes to respondents better.

Fourth, for a DCE study, the selected attributes should be exhaustive, covering all important and major aspects regarding a good or service of interest and reflecting the range of situations that respondents might expect to experience (Kjaer, 2005). While it is not feasible to include all relevant attributes in a single DCE design, researchers should put efforts into mimicking a real-world decision-making scenario for target respondents in their DCE questionnaires and include the most important attributes to respondents to describe the product or service under evaluation. For instance, if researchers want to investigate individuals' job preferences, *salary* should be selected as one of the attributes as it is a major consideration and influential factor for most job seekers. The exclusion of important attributes will likely result in biased results and estimates.

Finally, the selected attributes should be objectively measurable and actionable (Hensher et al., 2005). Attributes could be *quantitative* (e.g., salary, compensation, time, distance, and the number of treatments or *qualitative* (e.g., quality of services, providers of care, types of leisure activities). Quantitative attributes can be easily interpreted, but researchers need to pay extra attention when defining qualitative attributes and deciding on qualitative scales to use (ordinal or categorical) (Ryan et al., 2008). Importantly, attributes should be experimentally manipulable and defined in a way that gives room for making trade-offs between different attributes and attribute levels (Abiiro et al., 2014; Hensher et al., 2005). Researchers should also avoid situations where respondents rely on subjective evaluations of the attributes. Hence, an attribute should not be intrinsic or describes characteristics of the respondents themselves (e.g., personality, personal behaviors) rather than of the product, service, policy, or treatment under examination. In some cases, the "person" might be the target of DCE evaluations. For example, we can use DCEs to elicit individual preference toward political leaders and understand the relative importance of different attributes related to political leadership, such as honesty, charisma, policy expertise, and political ideology.

Besides these four important characteristics, researchers should also take into account the possible *interaction effects* between two or more attributes when selecting DCE attributes (also known as the "multinomial interactions" in the context of DCE studies). According to Ryan and colleagues (2008), interaction effects refer to situations where the *utility of an attribute is largely dependent on the levels of one or more other attributes*, such as price and quality. In many studies, two attributes are often *mutually dependent* as one increases, the other one will also increase. Another example is the screen size of a cell phone and its weight. Normally, a cell phone with a larger screen size will weigh more and be less portable than those with smaller screens. Researchers could address this situation by combining the two attributes into one or defining one of the attributes in the introductory text at the beginning of the DCE questionnaire and only including one in the choice sets (Kjaer, 2005).

In addition, researchers should also take into account the cognitive burden that may be imposed on respondents by including a large number of attributes or levels. Therefore, it is crucial to strike a balance between having enough attributes and levels to capture the complexity of the decision-making scenario and not overwhelming respondents with too much information.

Overall, appropriately selecting and defining DCE attributes is a critical step in the design to avoid biased or inaccurate preference estimates. The chosen attributes should be *policy-related, demand-related, measurable, and actionable* (Abiiro et al., 2014; Coast et al., 2012; Kjaer, 2005).

6.1.2 Approaches for Identifying DCE Attributes

In general, a *parsimonious design* is desirable. That is, a DCE study should have a limited selection of attributes and attribute levels is to make the study feasible and manageable for researchers and respondents (Pérez-Troncoso, 2020). This approach can help ensure that the study is not overly complex, which can lead to difficulties in data analysis and interpretation, as well as respondent fatigue and lack of engagement. A parsimonious design can also help ensure that the study is cost-effective, as it minimizes the resources needed to conduct the study.

Although there is no restriction on the number of attributes that could be included in a DCE, a *rule of thumb* is that a DCE study should contain *fewer than ten attributes* to ensure that respondents are able to consider all attributes listed when making decisions (DeShazo & Fermo, 2002). As the number of attributes in a DCE study increases, the cognitive burden on respondents increases, making it more challenging to complete the questionnaire accurately and efficiently (Kløjgaard et al., 2012). According to a literature review on the applications of DCEs in health economics by Soekhai and colleagues (2019), the number of attributes in most DCE studies ranged between *four to nine*.

A common strategy for establishing DCE attributes is first to *identify a long list of potential and relevant attributes and then scale down and reduce* the number of attributes to a number manageable within a DCE study (Abiiro et al., 2014). While DCE is a quantitative method to estimate individual preference, a qualitative approach is often used to identify DCE attributes and levels (Coast & Horrocks, 2007). As a general principle, the selection of attributes and attribute levels should be guided by systematic approaches or justifiable by evidence (Pérez-Troncoso, 2020). However, existing DCE studies shed little light on how qualitative work should be translated into the development of attribute and attribute levels in DCEs.

We provide below the common approaches for identifying and developing DCE attributes: (1) literature review, (2) interviews, (3) expert opinion, and (4) social media analysis. We will introduce these approaches one by one and discuss their applications in existing DCE studies.

6.1.2.1 Literature Review

A literature review is perhaps one of the most commonly used approaches for identifying DCE attributes. A rigorous and systematic literature review on the topics of research interest from literature databases such as Google Scholar, ScienceDirect, Web of Science, EBSCOhost, and PubMed could help researchers identify a comprehensive list of conceptual attributes to be included in a DCE study (Abiiro et al., 2014). Some studies relied solely on a literature review to derive attributes to be included in DCE studies (e.g., Kamphuis et al., 2015; Scott, 2002; van Rijnsoever et al., 2015). We can also use published papers, reviews, and policy documents to help identify attributes and levels. For example, Hazlewood and colleagues (2016, p. 1960) reported their attribute development process based on a literature review:

> *One author generated a list of potential attributes from outcomes evaluated in Cochrane reviews and a literature review that included a grey-literature search of printed patient material and decision aids based on Cochrane reviews...the final attributes included three trial outcomes (major symptom improvement, joint damage, stopping due to a side effect), the dosing regime, rare adverse events and required monitoring or lifestyle restrictions (e.g., alcohol use).*

Another example is a DCE study that examines public preference toward energy technologies by van Rijnsoever et al. (2015, p. 821):

> *The values of the levels for each technology were based on a review of scientific and policy documents and verified by academic experts in the field of energy research.*

While it is much easier and more convenient for researchers to identify attributes exclusively on the basis of a literature review, it is important to note that relying solely on a literature review may not capture the full range of important attributes and levels that are relevant to the target population. It could potentially lead to the *non-inclusion* of some important attributes, especially those *context-specific* attributes (Coast et al., 2012). Hence, many studies *combined primary and secondary data* in identifying attributes. This means performing a literature review as the initial stage of attribute development, followed by primary data such as qualitative research involving consulting experts or conducting focus group discussions (e.g., Bronnmann & Asche, 2017; van Empel et al., 2011). One example is the study by Marti (2012, p. 536):

> *From existing literature, we initially developed a list of attributes about smoking cessation pharmaceutical treatments that seemed the most relevant to our research goals... Meanwhile, we conducted two focus groups with five participants in order to identify important attributes and prevent the omission of salient ones.*

Similarly, a study by Czoli and colleagues (2015, p. e31) combined literature review and consultation with experts to identify important, policy-relevant, and context-specific attributes in their DCE study:

> *Product attributes and their corresponding levels were selected on the basis of a review of the literature and through consultation with experts in e-cigarette and cigarette product science and policy to ensure that they accurately reflected the market at the time of the study.*

6.1.2.2 Interviews

Conducting interviews is a very useful approach for identifying context-specific attributes in DCE studies. Interviews can capture deep insights and feedback from relevant stakeholders or target populations, increasing the *face validity* of DCEs and ensuring that the DCE is tailored to the study setting and contexts (Abiiro et al., 2014; Kløjgaard et al., 2012). Interviews could be conducted in various ways, such as *in-depth interviews, semi-structured interviews, and focus group interviews* with samples of relevant respondents, experts, and/or other stakeholders. For example, Bekker-Grob and colleagues (2010b) conducted a DCE study to examine girls' preference for HPV vaccination. To identify the most relevant HPV vaccination attributes, the researchers performed focus groups with 36 parents:

> *In the focus groups we collected data on the attributes that individuals expected to be important or that had been important in their decision to participate in an HPV vaccination programme.* (Bekker-Grob et al., 2010a, p. 6693)

Researchers should recruit interviewees with intimate knowledge or personal or professional experiences with the product, service, or idea under investigation. For example, Mandrik and colleagues (2019) conducted a DCE study investigating population preference for breast cancer screening. To test the importance, relevance, and clarity of attributes identified from their initial literature review, they conducted in-depth interviews with *three categories of key informants*, including healthcare professionals of the screening centers, health professionals who are not involved in screening, and the target population of the breast cancer screening (i.e., 50–69-year-old women who participated and did not participate in screening mammography).

Importantly, interviews should be guided by an *interview protocol* (i.e., a list of pre-designed questions), and the interview contents should focus on exploring participants' perceptions, beliefs, attitudes, and experiences. An interview protocol, which could be developed from the results of the initial literature review (Bottomley et al., 2017), will help researchers have a clear focus in their interviews and consistency in how questions are asked. As reported in a DCE study by Bottomley and colleagues (2017, p. 864):

> *A semi-structured interview guide was developed from the results of the literature review, and telephone, web-assisted interviews were conducted with 12 people with MS, recruited through a specialist medical recruitment agency. Participants were asked to first spontaneously report the treatment attributes which were of most and least importance to them, and then review the list of potential study attributes.*

6.1.2.3 Expert Opinion

Policy and practical concerns may shape the choice of attributes. Our selected attributes should be actionable, and the results of DCE studies should offer practical implications to improve existing products, services, or policies (Kløjgaard et al., 2012). Therefore, it is often suggested to *engage professional practitioners, policy-makers, and/ or local institutions* during the development of DCE attributes. Expert opinions could be collected and utilized in various ways. Researchers can conduct qualitative interviews with experts who are knowledgeable and/or very experienced with the topic under investigation to determine attributes that are of great policy interest or have significant practical values (Abiiro et al., 2014). For instance, a DCE study by Günther and colleagues (2010) was preceded by interviews with 22 physicians to identify relevant attributes regarding their preference for practice establishment in Germany, as the authors illustrated:

> *Attributes relevant to the establishment of a practice were identified, described and labeled to ensure the BWS task encompassed the most relevant attributes by executing a qualitative study interviewing 22 physician in-depth... The interviewed physicians had not yet established a practice but intended to.* (p. 216)

Another example is a DCE study examining indigenous fishermen's preference and willingness to pay values for bequest gains from management actions in a marine area. The authors developed and refined attributes and attribute levels development using expert consultations, as they illustrated (Oleson et al., 2015):

> *Attributes and levels were initially developed during focus groups, and were refined after several expert consultations with local collaborators and key informants, and after a pre-test in multiple villages. All levels either reflected current conditions or experts' opinions on expected changes due to management or lack thereof*. (p. 107)

Moreover, expert opinions can also be used to *reduce the number of total attributes* (i.e., dimension reduction) to be included in the final DCE design. In some DCE studies, experts are asked to *give feedback* on the concepts identified in the literature, commenting on the importance of each concept. Researchers may ask experts to use a scale (e.g., a 5-point Likert scale) to rate a list of potential attributes based on their relevance and importance, as in the study by Karyani and colleagues (2018):

> *In the third phase, a candidate list of attributes from the previous steps was prepared...Therefore, these attributes and attributes levels were prioritized and rated by the opinion of 36 experts. Each attribute was rated using a scale of 1 to 5, with 1 indicating the lowest score and 5 the highest score. The scores of each feature were summed up to obtain an overall score and rank.* (p. 2)

6.1.2.4 Social Media Analysis

Finally, social media analysis can also be used as an alternative approach for identifying and developing attributes for DCE studies. Researchers often use *keywords* or *hashtags* related to their research topics to search on social media platforms (e.g.,

Twitter, Facebook, Instagram, LinkedIn) to learn about how the topic of interest was discussed in the public domain. For example, if researchers are interested in understanding consumer preferences for sustainable food products, they may search for keywords such as "sustainable food", "organic food", or "eco-friendly food".

This approach can yield large amounts of data on how people discuss and think about these topics. By using social media analysis to identify and develop attributes for DCE studies, researchers can gain valuable insights into how people perceive and make decisions about the products, services, or policies under investigation. This approach can also provide a more diverse and representative sample of respondents, as people from a wide range of backgrounds and demographics use social media platforms.

For instance, a DCE study by Barber and colleagues (2019) performed a systematic analysis on a number of 176 social media posts relating to hypodontia to identify attributes related to their study, as they reported:

> *To provide additional information from the perspective of people affected by hypodontia, a systematic search and evaluation of posts relating to hypodontia on social media was performed.* (p. 140)

In addition, Genie and co-authors (2020) conducted a DCE study to examine public preference for government responses during a pandemic. They first selected potential attributes to be included in their DCE design based on policy discussions and government guidance in European countries. To understand how these attributes were discussed by the public, the authors extracted 15,000 tweets that contained *phrases or words* related to different potential attributes (e.g., types of lockdown, lockdown length) and performed a *sentiment analysis* to categorize public opinions in the text related to their attributes, as they illustrated below (Genie et al., 2020):

> *Then we used social media analysis to gain insight into how these attributes were discussed in the public domain. We conducted localised searches for tweets that contained phrases or words that could be used to describe the attributes. We generated a sentiment analysis29 30 from the tweets to illustrate how people were construing these words when related to the attributes. This provided insight as to what was important to the general public and how it was being talked about.* (p. 4)

This section briefly discussed four commonly used approaches for identifying and developing DCE attributes. Each approach has its own strengths and limitations. For instance, focus group interviews with target respondents could provide valuable information on how a specific product or service is perceived by populations and identify attributes that are most relevant and context-specific to the local populations. Having consultations with experts or key specialists could provide useful insights for attribute selection and enhance the realism and plausibility of the DCE study (Ryan et al., 2008). Given that poorly selected attributes will eventually jeopardize the final results of a DCE study (Pérez-Troncoso, 2020), we recommend that researchers combine different approaches and combine primary and secondary data to increase the validity and reliability of the attribute identification process.

6.1.3 Establishing Attribute Levels

After the attributes are selected and defined, *attribute levels* need to be assigned. Assigning appropriate levels to DCE attributes will increase the precision of parameter estimates in a DCE study (Hall et al., 2004). Generally, the levels of each attribute should be realistic and reflect the range of situations that target populations might expect to encounter in the real world (Kløjgaard et al., 2012). A good DCE design should have policy or practically relevant and plausible attribute levels with *enough variation* to produce meaningful behavioral responses from respondents (Ryan et al., 2008). In other words, the attribute levels should be constructed so that respondents are willing to *make trade-offs* between alternatives (Kjaer, 2005). If attribute levels are too narrow or too wide, respondents may consider the difference between levels to be insignificant or too significant, which will ultimately result in dominating levels and non-trading responses (Green & Srinivasan, 1990).

Moreover, the number of attribute levels also influences data analysis and results of a DCE study. As Ryan et al. (2008) demonstrated when only two levels are assigned for a DCE attribute, the marginal utility function of the attribute can only be linear. An increase in the number of levels assigned will allow researchers to *detect more complex, non-linear utility* relationships between different levels. By increasing the number of levels, the researcher can better capture the range of variation in the attributes being tested, making it easier to identify how changes in those attributes impact the decision-making process. However, while more levels could provide a better understanding of the utility that respondents place on an attribute, too many levels will inevitably increase the complexity of the DCE tasks and increase the cognitive difficulty for respondents to complete the questionnaire. Obadha and collaborators (2019) suggested that a DCE study should restrict the maximum number of levels of each attribute to four to ensure the overall experiment is deemed manageable for the respondents.

It is often not easy for researchers to determine which attribute levels are acceptable to target respondents. We address this challenge using two approaches.

First, for establishing attribute levels, researchers could make references to the status quo by investigating those products or services available on the market or referring to previous administrative and testing records, as illustrated by Hol and colleagues in a DCE study (2010):

> *The specific values (levels; e.g. amount of risk reduction or length of screening interval) for each test characteristic incorporated the range of possible test outcomes of a specific screening test based on the current literature. The levels were test-specific to create realistic scenarios.* (p. 973)

Similarly, Van Empel and colleagues (2011) made reference to existing clinical records when assigning attribute levels in their DCE study:

> *Each attribute was divided into three meaningful levels that covered the 'realistic range'… For example, the levels of 'pregnancy rate' ranged from 20 to 35%, which is consistent with the range of the mean ongoing pregnancy rate per IVF-cycle in Dutch and Belgian fertility clinics.* (p. 585)

Second, researchers could also ask target respondents to identify probable options or values for each attribute using interviews or pilot tests for deriving specific attribute levels. This approach is particularly useful as it helps derive attribute levels that are considered important and meaningful to respondents and will ensure that the attribute levels will be significant enough for respondents to make trade-offs. For instance, Alten and colleagues (2016) conducted focus group discussions to establish both attributes and their levels, as they illustrated:

> *Discussions with focus groups were conducted to identify relevant DMARD attributes and their levels to be used in the final DCE. An initial list of attributes and their levels ... was presented to participants at the outset of discussion. After discussion on why these attributes felt important and in how far their different levels presented a perceivable and meaningful difference, each patient was asked to rank attributes in order of their subjective importance.* (p. 2219)

Another example is a DCE study by Masiero and colleagues (2015) that examined guests' willingness to pay for hotel room attributes. They reported the process of establishing attribute levels that include discussion with business managers, as shown below:

> *In the first stage of the study, a meeting with hotel managers was arranged in order to better identify the attributes and attribute levels to be used in the stated choice experiment so that the study could be beneficial to the property. After the meeting and extensive discussion among the research team, seven attributes with different levels were identified.* (p. 119)

Establishing attribute levels is an important but challenging step when designing a DCE study. Norman and colleagues (2020) proposed three criteria researchers can follow to develop DCE attribute levels. These criteria are: (1) all levels should be plausible, (2) all levels should be similar to current practice in the study setting (context-relevant), and (3) levels should be spread enough to allow respondents to make a trade-off between them, but not too spread so that an option with the poorer level of an attribute will never be picked. We will use our own research project in the section below to demonstrate the process of developing DCE attributes and attribute levels.

6.1.4 An Example of Establishing DCE Attributes and Levels

As mentioned in Chap. 5, we conducted a real research project entitled "*Examining Consumer Preference for Alternative Meat*", which we use as an illustration in this book. We offer step-by-step instructions for designing and implementing a DCE study and performing data analysis. In this demonstrative project, we are interested in examining consumer *preference for alternative meat products*, which are plant-based food products made entirely from plants or vegan ingredients. Alternative meat products aim to mimic the appearance, texture, and taste of real (animal-based) meat, and they can be found in various products such as patties, cold cuts, and sausages.

One example of alternative meat is the Beyond Meat burger patty meat, which is entirely made of plants (see Fig. 6.1).

This demonstrative project aims to explore how consumers perceive alternative meat products and their preferences when purchasing. To establish attribute and attribute levels, we followed a *standard iterative process* (Liao et al., 2020; Ryan et al., 2007) by adopting a three-step process that combines *literature review, analysis of online discussions, and focus group interviews*. We believe that integrating secondary and primary data will allow us to better identify appropriate attributes to be included in the DCE design and ultimately increase the validity and accuracy of the study.

For this demonstrative project, we started the attribute development with an initial literature review of articles extracted from Google Scholar and Web of Science. We used the keywords "*alternative meat*", "*plant-based meat*", "*meat alternatives*", and "*vegetarian meat*" to search for relevant academic articles. We selected articles that discussed the emergence and significance of alternative meat products and consumers' attitudes toward them. A total of 13 articles were included in our initial literature review to explore important factors related to alternative meats (see Table 6.1 for examples of our reviewed articles and identified factors).

Three members of our research team completed the literature review, and all the team members discussed the findings together. In particular, our literature review

Fig. 6.1 An example of an alternative meat product

Table 6.1 Examples of initial literature review in DCE studies

Academic articles	Key factors	Sample quotes
Slade (2018)	Environmental Protection	*Environmental Impact is the most significant factor in plant-based and cultured meat purchase behaviour*
	Price	*If the price were equal, 65% of consumers would buy the beef burger rather than plant-based(21%) or cultured meat (11%)*
	Health	*Health is the third significant factor in plant-based meat purchase behaviour*
	Naturality	*Participants who preferred natural food would consider cultured meat is unnatural, they are less likely to buy cultured meat*
Thavamani et al. (2020)	Ethical Considerations	*Vegetarians have lesser acceptance on cultured meat because cultured meat is derived from animals*
	Environmental Impact	*Consumers may be attracted to meat alternatives for reasons, including the sustainability and impact on the environment*
	Vegetarian Diet	*Participants considered the character and attributes of the meat alternatives as the vital factors that influence their preferences other than the impact on the environment and health*
	Nutritional Values	*Nutritionists stated that plant-based meat is oftern over-process and have reduced nutritional value of ingredients. Most of the plan-based meat have low saturated fat but have increased high sodium content*
Estell et al. (2021)	Environmental Protection	*Around 67% of consumers considered plant-based meat alternatives are more environmentally friendly than meat*
	Curiosity	*Curiosity to try new foods was the top reason for the consumers to try Plant-based meat alternatives*
	Price	*Cost is anotther factor influencing purchasing decisions*
Arora et al. (2020)	Environmental Protection	*Animal-based proteins produce higher emissions and require more land than vegetal sources of protein*
	Health	*The health benefits of plant-based diets have been widely documented, too*
	Price	*Alternatively, we likewise estimate that the price of plant-based meat would need to decrease by 65%, and the price of clean meat would need to decrease by 95%, to increase their respective market shares by 50%*

(continued)

Table 6.1 (continued)

Academic articles	Key factors	Sample quotes
Escribano et al. (2021)	Production technology	*Type of production seems to be one of the most significant attributes shaping consumer preferences, and due to the novelty of both plant-based and cultured meats, this could be one of the main reasons for consumer rejection*
	Environmental Protection (carbon footprint)	*Indeed, some consumers (sustainable consumers) placed great significance on the labels, although in the study not even the "Conscious/concerned consumers" displayed this behavior*
	Price	*Most of members of "Price-sensitive millennials", who were under 30 years old placed the greatest importance on price, which may be related to their lower income levels*

identified five factors that may affect the public's willingness and attitude when considering alternative meat products: environmental protection, ethical consideration, nutritional values, health concerns, and costs. Each attribute is briefly discussed in Textbox 6.1.

Textbox 6.1. Factors Related to Alternative Meats from Literature

Environmental Protection

- Compared to conventional meat products, the production of alternative meats causes much less greenhouse gas emissions
- The production of alternative meats could reduce energy consumption, land usage, and water usage significantly
- The consumption of alternative meats promotes vegan and flexitarian diets, which could reduce 70% and 40% of greenhouse gas emissions

Ethical Consideration

- Alternative meat products are entirely plant-based and thus address public's ethical concerns about killing animals
- Reducing animal deaths and minimizing animal cruelty by replacing conventional meat consumption with alternative meat products

Nutritional Values

- Most alternative meat products are made of soy, which is a good source of protein and fibre
- Other plant-based ingredients are also a source of vitamins, minerals, and antioxidants, which have health benefits to consumers

- Alternative meats typically are lower in calories compared to conventional meats

Health Concerns

- Some alternative meat products contain additives and artificial fillers that are high in sodium and thus not healthy to consumers

Costs

- The production cost of alternative meat is relatively high. It could cost up to US$8 per pound

After getting some initial insights from the literature review, we further collected secondary data from *online forums, blogs, and media reports* to understand how the public discussed alternative meat products. The research team coded the public's comments and opinions regarding alternative meat products and *categorized* them into two major groups: *Positive and Negative.* Below we provide some examples of our coding and positive and negative comments from citizens (see Textboxes 6.2 and 6.3).

Textbox 6.2. Positive Comments related to Alternative Meats

1. "I am a vegetarian. I was unpleasant to see news reporting sick pigs were killed inhumanely. I recommend everyone can try Omni Pork to replace pork. I think it is healthy, tasty and ethical" *(ethical consideration)*
2. "From environmental/ economic perspective, the carbon dioxide and other gases emitted by meat farms are contributing to global warming and changing land use and using large amounts of water. These are ecologically damaging. It is better to provide more choice to the consumer market, which reduces killing, is a good thing" (*environmental protection/ economic efficiency*)
3. "I'm a vegetarian, and I think it's nice to remember the taste of lunch meat. It's good for us vegetarians to have another option. It's so nice to taste it without killing anything" (*ethical consideration/ taste*)
4. "Does anyone eat new pork? (Omni Pork, you can search it online). It looks like pork, but the texture is sticky, and the taste is not like meat. Eating healthy and less fat is the most important thing for me to lose weight" (*health*)
5. "Omnipork mince is delicious when it is fried with tofu. It looks like real meat. I didn't taste the difference between the Omnipork mince and real meat for the first trial" (*taste*).

Textbox 6.3. Negative Comments related to Alternative Meats

1. "Vegetarians are not interested in alternative meet. Alternative meet is made of vegetable protein, which is converted into animal-like protein. If you are vegans, your stomach will not be feeling well (diarrhoea or vomit) after eating alternative meet." *(vegan/ health issues)*
2. "Omnipork tastes like dried bean curd, I don't like it. And I think it may not be healthier than (real) pork because of the seasoning" (*taste poorly*)
3. "The ingredients are not healthy; it mainly uses soy protein; it may be not good to eat them for a long term. Alternative meat will use MSG as seasoning and add synthetic vitamin." (*health issues*)
4. "Don't forget that these human-made (artificial) alternatives meats are expensive than real meat" (*cost*)
5. "Nutritionists say artificial meat is a processed food with preservatives and additives. It is not necessarily a healthy food. In fact, it is not necessary to be completely vegan without religious belief or physical necessity. It is better to eat more vegetables and less meat, and to eat less processed food" (*health issues*).

The secondary data analysis collected from public discussions online further confirmed the factors identified in our initial literature review by demonstrating that *environmental protection, ethical consideration, nutritional values, health concerns, and costs* could shape the public's attitudes toward alternative meat products. Besides, we also found that the *taste* of alternative meats (e.g., whether it tastes like real meat or not and whether it is well-seasoned) may also affect the public's willingness to purchase and try alternative meat products.

On the basis of the findings from the literature review and analysis of online discussions, we subsequently conducted *two focus group interviews* with four participants in each group (8 participants in total) as a sample of general consumers to further understand their views and perspectives about the important attributes of alternative meat products and identify reasonable levels for each attribute. Since our study deals with a *non-sensitive topic*, focus group interviews were considered appropriate for identifying attributes from general populations as they could quickly yield large amounts of consensual information on a specific topic (Louviere et al., 2000). We created an interview protocol to guide our interviews (see Textbox 6.4).

Textbox 6.4. Interview Guide for Focus Group Interviews

General Questions

- Have you tried any meat alternative meat products?
- Do you know the name of the brands of the alternative meat?
- Did you buy the alternative meat from supermarkets or restaurants? Which one?
- How long ago did you last buy/eat alternative meat?
- Are you a regular eater of alternative meat or just try it once?

- What does Alternative Meat mean to you?

Experience with Alternative Meat

- There is plenty of real meat that you can buy easily, but why did you try alternative meat?
- How did you feel after you ate an alternative meat product? What was the experience like compared to real meat?
- In your experience, is alternative meat different from vegetarian products? How different?
- What factors do you consider when purchasing an alternative meat product? (if no specific answers from participants, *probe these: taste, color, smell, texture, nutrients, price, availability in stores, recommendation by friends/families, advertisement*)
- Does alternative meat have advantages compared to conventional meat? Explain
- Does alternative meat have disadvantages compared to conventional meat? Explain
- Does alternative meat have advantages compared to vegetarian products? Explain
- Does alternative meat have disadvantages compared to vegetarian products? Explain
- What are the most important factors for you when choosing an alternative meat?

Future Behavior

- Will you continue to purchase and consume alternative meat in the future?
- Will you recommend alternative meat to friends and family members?

We recruited interview participants using *purposive sampling* by selecting those individuals who *have tried alternative meats* before and/or have a basic understanding of alternative meat. Table 6.2 summarizes the demographic information of participants included in our focus group interviews.

All focus group interviews were conducted by the research team and were tape-recorded and transcribed for analysis. We *coded* interview transcripts by categorizing interviewees' comments and views on alternative meat products into *Positive and Negative* and identified important attributes that interviewees repeatedly mentioned to understand likely causes of positive and negative attitudes toward alternative meat products (see Table 6.3 for a sample of our data analysis).

Consequently, combining the insights that we gained from all three phases (i.e., literature review, analysis of online discussions, and focus group interviews), we *determined 6 attributes* that describe characteristics of alternative meat products

Table 6.2 Demographic information of interview participants

Interviewee	Age group	Education level	Monthly income (US$)	Marital status	Employment status
1	40–44	Bachelor's Degree	4,400–5,000	Married	Full-Time
2	> 50	Diploma	< 1,200	Single	Part-Time
3	30–34	Postgraduate Education	1,200–1,800	Single	Part-Time
4	20–24	Bachelor's Degree	< 1,200	Single	Freelancer
5	20–24	Postgraduate Education	2,501–3,200	Single	Full-Time
6	20–24	Bachelor's Degree	1,200–1,800	Single	Full-Time
7	40–44	Secondary Education	1,801–2,500	Single	Full-Time
8	30–34	Bachelor's Degree	4,400–5,000	Single	Full-Time

Table 6.3 A sample of analyses of interviews

Interview data analysis	
Positive	
Taste	*I have tried a type of planted-based patty, which tasted like a real meat patty. It seemed saltier than real patty*
	Although the texture is not as soft as the real nuggets, the taste is quite good
Environmental Protection	*I think if people's meal habits are changed to meat alternative meals, that would reduce the demand for real meat*
	I think meat alternative could reduce carbon footprints, which is a good thing for the whole environment
Health	*The calories of meat alternatives are lower than real meat*
	I think alternative meat was healthy for my stomach. Real meat makes me feel serious flatulence. I think meat alternative is an digestible food
Negative	
Additives	*It shows that there are many additive, preservative. I also want to eat healthier but there are many*
	There are too much additive. The reason for me to eat plant-based meat is because there are too many additive, medicine and hormone injection
Price	*I think alternative meat is more costly than real meat*
	I think being costly is the disadvantage
Texture	*I think cookery affects the taste of alternative meat products. For instance, the taste of Omipork with soup noodles was bad. It tasted like pork chop with a lot of meat tenderizer. The texture was unreal*
	Some products are frozen, I think they are not as convenient as packaged food

from different aspects, which are *additives, nutrition, format, vegan label, carbon footprint, and price.* We defined and explained each attribute in detail in Textbox 6.5.

Textbox 6.5. Defining Attributes

1. **Additives**: referring to whether additives are used to improve the product's flavor or appearance, or to keep it fresh or to prevent it from decay.
2. **Nutrition**: referring to the "good and healthy thing" that you put in your body and we use "protein" as the definition for nutrition.
3. **Format**: referring to whether the product requires any preparation before consumption, or whether it can be consumed right away.
4. **Vegan Label**: referring to a label that indicates whether the product contains any real meat or not. Real meat includes anything from red meat, poultry, seafood, or any parts of an animal.
5. **Carbon Footprint**: the overall amount of greenhouse gases produced by a product, such as fossil fuel used to transport products, the pollution from growing the animal, etc.
6. **Price**: referring to the amount of money that one must pay to buy the product.

For establishing attribute levels, we selected the range of protein amount, carbon footprint emission levels, and prices to be in line with the options available on the local market at the time of the study. To ensure the final DCE questionnaire is manageable to respondents, each attribute is assigned two or three levels only. Table 6.4 provides an overview of attributes and levels to be included in this sample DCE study.

Table 6.4 Final attributes and corresponding levels

Attributes	Attribute levels (a 200-g of alternative meat)
Additives	(1) Without additives
	(2) With additives
Nutrition	(1) 25 g of proteins
	(2) 40 g of proteins
Format	(1) Raw
	(2) Ready-to-eat
Vegan Label	(1) Without vegan label
	(2) With vegan label
Carbon Footprint	(1) No information on reduction in greenhouse emission
	(2) 20% of reduction in greenhouse gas emission
Price	(1) US$3.5
	(2) US$4.5
	(3) US$6.0

After deciding the attributes and attribute levels to be used in the DCE, researchers could then proceed to the experimental design. In the next chapter, we will discuss several principles of DCE design and provide step-by-step instructions on utilizing R programming language to help construct DCE questionnaires.

References

Abiiro, G. A., Leppert, G., Mbera, G. B., Robyn, P. J., & De Allegri, M. (2014). Developing attributes and attribute-levels for a discrete choice experiment on micro health insurance in rural Malawi. *BMC Health Services Research, 14*(1), 1–15.

Alten, R., Krüger, K., Rellecke, J., Schiffner-Rohe, J., Behmer, O., Schiffhorst, G., & Nolting, H. D. (2016). Examining patient preferences in the treatment of rheumatoid arthritis using a discrete-choice approach. *Patient Preference and Adherence, 10*, 2217–2228.

Arora, R. S., Brent, D. A., & Jaenicke, E. C. (2020). Is India ready for alt-meat? Preference and willingness to pay for meat alternatives. *Sustainability, 12*(11), 4377.

Barber, S., Bekker, H., Marti, J., Pavitt, S., Khambay, B., & Meads, D. (2019). Development of a discrete-choice experiment (DCE) to elicit adolescent and parent preference for hypodontia treatment. *The Patient-Patient-Centered Outcomes Research, 12*(1), 137–148.

Bennett, J., & Blamey, R. (Eds.). (2001). *The choice modelling approach to environmental valuation.* Edward Elgar Publishing.

Bottomley, C., Lloyd, A., Bennett, G., & Adlard, N. (2017). A discrete choice experiment to determine UK patient preference for attributes of disease modifying treatments in Multiple Sclerosis. *Journal of Medical Economics, 20*(8), 863–870.

Bronnmann, J., & Asche, F. (2017). Sustainable seafood from aquaculture and wild fisheries: Insights from a discrete choice experiment in Germany. *Ecological Economics, 142*, 113–119.

Coast, J., & Horrocks, S. (2007). Developing attributes and levels for discrete choice experiments using qualitative methods. *Journal of Health Services Research & Policy, 12*(1), 25–30.

Coast, J., Al-Janabi, H., Sutton, E. J., Horrocks, S. A., Vosper, A. J., Swancutt, D. R., & Flynn, T. N. (2012). Using qualitative methods for attribute development for discrete choice experiments: Issues and recommendations. *Health Economics, 21*(6), 730–741.

de Bekker-Grob, E. W., Hofman, R., Donkers, B., van Ballegooijen, M., Helmerhorst, T. J., Raat, H., & Korfage, I. J. (2010a). Girls' preference for HPV vaccination: A discrete choice experiment. *Vaccine, 28*(41), 6692–6697.

de Bekker-Grob, E. W., Hol, L., Donkers, B., Van Dam, L., Habbema, J. D. F., Van Leerdam, M. E., … & Steyerberg, E. W. (2010b). Labeled versus unlabeled discrete choice experiments in health economics: An application to colorectal cancer screening. *Value in Health, 13*(2), 315–323.

DeShazo, J. R., & Fermo, G. (2002). Designing choice sets for stated preference methods: The effects of complexity on choice consistency. *Journal of Environmental Economics and Management, 44*(1), 123–143.

Escribano, A. J., Peña, M. B., Díaz-Caro, C., Elghannam, A., Crespo-Cebada, E., & Mesías, F. J. (2021). Stated preference for plant-based and cultured meat: A choice experiment study of Spanish consumers. *Sustainability, 13*(15), 8235.

Estell, M., Hughes, J., & Grafenauer, S. (2021). Plant protein and plant-based meat alternatives: Consumer and nutrition professional attitudes and perceptions. *Sustainability, 13*(3), 1478.

Genie, M. G., Loría-Rebolledo, L. E., Paranjothy, S., Powell, D., Ryan, M., Sakowsky, R. A., & Watson, V. (2020). Understanding public preference and trade-offs for government responses during a pandemic: A protocol for a discrete choice experiment in the UK. *British Medical Journal Open, 10*(11), e043477.

Green, P. E., & Srinivasan, V. (1990). Conjoint analysis in marketing: New developments with implications for research and practice. *Journal of Marketing, 54*(4), 3–19.

Günther, O. H., Kürstein, B., Riedel-Heller, S. G., & König, H. H. (2010). The role of monetary and nonmonetary incentives on the choice of practice establishment: A stated preference study of young physicians in Germany. *Health Services Research, 45*(1), 212–229.

Hall, J., Viney, R., Haas, M., & Louviere, J. (2004). Using stated preference discrete choice modeling to evaluate health care programs. *Journal of Business Research, 57*(9), 1026–1032.

Hazlewood, G. S., Bombardier, C., Tomlinson, G., Thorne, C., Bykerk, V. P., Thompson, A., ... & Marshall, D. A. (2016). Treatment preference of patients with early rheumatoid arthritis: A discrete-choice experiment. *Rheumatology, 55*(11), 1959–1968.

Hensher, D. A., Rose, J. M., Rose, J. M., & Greene, W. H. (2005). *Applied choice analysis: A primer*. Cambridge University Press.

Hol, L., de Bekker-Grob, E. W., van Dam, L., Donkers, B., Kuipers, E. J., Habbema, J. D. F., ... & Essink-Bot, M. L. (2010). Preference for colorectal cancer screening strategies: a discrete choice experiment. *British Journal of Cancer, 102*(6), 972–980.

Kamphuis, C. B., de Bekker-Grob, E. W., & van Lenthe, F. J. (2015). Factors affecting food choices of older adults from high and low socioeconomic groups: A discrete choice experiment. *The American Journal of Clinical Nutrition, 101*(4), 768–774.

Karyani, A. K., Rashidian, A., Sari, A. A., & Sefiddashti, S. E. (2018). Developing attributes and levels for a discrete choice experiment on basic health insurance in Iran. *Medical Journal of the Islamic Republic of Iran, 32*, 26.

Kjaer, T. (2005). *A review of the discrete choice experiment-with emphasis on its application in health care*. Syddansk Universitet.

Kløjgaard, M. E., Bech, M., & Søgaard, R. (2012). Designing a stated choice experiment: The value of a qualitative process. *Journal of Choice Modelling, 5*(2), 1–18.

Liao, Q., Ng, T. W., & Cowling, B. J. (2020). What influenza vaccination programmes are preferred by healthcare personnel? A discrete choice experiment. *Vaccine, 38*(29), 4557–4563.

Louviere, J. J., Hensher, D. A., & Swait, J. D. (2000). *Stated choice methods: Analysis and applications*. Cambridge University Press.

Mandrik, O., Yaumenenka, A., Herrero, R., & Jonker, M. F. (2019). Population preference for breast cancer screening policies: Discrete choice experiment in Belarus. *PLoS ONE, 14*(11), e0224667.

Mangham, L. J., Hanson, K., & McPake, B. (2009). How to do (or not to do)… Designing a discrete choice experiment for application in a low-income country. *Health Policy and Planning, 24*(2), 151–158.

Marti, J. (2012). Assessing preference for improved smoking cessation medications: A discrete choice experiment. *The European Journal of Health Economics, 13*(5), 533–548.

Masiero, L., Heo, C. Y., & Pan, B. (2015). Determining guests' willingness to pay for hotel room attributes with a discrete choice model. *International Journal of Hospitality Management, 49*, 117–124.

Norman, R., Moorin, R., Maxwell, S., Robinson, S., & Brims, F. (2020). Public attitudes on lung cancer screening and radiation risk: A best-worst experiment. *Value in Health, 23*(4), 495–505.

Obadha, M., Barasa, E., Kazungu, J., Abiiro, G. A., & Chuma, J. (2019). Attribute development and level selection for a discrete choice experiment to elicit the preference of health care providers for capitation payment mechanism in Kenya. *Health Economics Review, 9*(1), 1–19.

Oleson, K. L., Barnes, M., Brander, L. M., Oliver, T. A., van Beek, I., Zafindrasilivonona, B., & van Beukering, P. (2015). Cultural bequest values for ecosystem service flows among indigenous fishers: A discrete choice experiment validated with mixed methods. *Ecological Economics, 114*, 104–116.

Pérez-Troncoso, D. (2020). A step-by-step guide to design, implement, and analyze a discrete choice experiment. *arXiv preprint*. arXiv:2009.11235.

Ryan, M., Gerard, K., & Amaya-Amaya, M. (Eds.). (2007). *Using discrete choice experiments to value health and health care* (Vol. 11). Springer Science & Business Media.

Ryan, M., Gerard, K., Watson, V., Street, D. J., & Burgess, L. (2008). Practical issues in conducting a discrete choice experiment. *Using Discrete Choice Experiments to Value Health and Health Care*, 73–97.

Scott, A. (2002). Identifying and analysing dominant preference in discrete choice experiments: An application in health care. *Journal of Economic Psychology, 23*(3), 383–398.
Slade, P. (2018). If you build it, will they eat it? Consumer preference for plant-based and cultured meat burgers. *Appetite, 125*, 428–437.
Soekhai, V., de Bekker-Grob, E. W., Ellis, A. R., & Vass, C. M. (2019). Discrete choice experiments in health economics: Past, present and future. *Pharmacoeconomics, 37*, 201–226.
Thavamani, A., Sferra, T. J., & Sankararaman, S. (2020). Meet the meat alternatives: The value of alternative protein sources. *Current Nutrition Reports, 9*(4), 346–355.
van Empel, I. W., Dancet, E. A., Koolman, X. H., Nelen, W. L., Stolk, E. A., Sermeus, W., ... & Kremer, J. A. (2011). Physicians underestimate the importance of patient-centredness to patients: a discrete choice experiment in fertility care. *Human Reproduction, 26*(3), 584–593.
van Rijnsoever, F. J., Van Mossel, A., & Broecks, K. P. (2015). Public acceptance of energy technologies: The effects of labeling, time, and heterogeneity in a discrete choice experiment. *Renewable and Sustainable Energy Reviews, 45*, 817–829.

Chapter 7
Designing DCE Choice Set Using R

The construction of experimental design is key to the successful implementation of DCE research. A *full factorial design* is the easiest design to create and use. It consists of all possible combinations of attribute levels, but it is often considered too costly and tedious for respondents. A recommended approach is to employ *fractional factorial design* in the DCE design, which uses a randomized but smaller combination of attribute levels. This chapter discusses techniques and steps to generate an *orthogonal* and *balanced fractional factorial* DCE design using R packages, such as package **support.CEs**.

7.1 Full Factorial and Fractional Factorial Designs

An important step in implementing a DCE study is constructing an efficient experimental design. As echoed by Weber (2021), an experimental design in DCE is fundamentally nothing more than a *matrix of values* that indicates what goes where. Table 7.1 presents an example of an experimental design in which every row represents a *choice task* (e.g., choose Alternative A or B or Neither), while every column represents an *attribute* of an alternative. The values (i.e., 0, 1, 2) represent the *attribute levels* (0 g vs. 10 g vs. 20 g of sugar) that we use in the experiment.

This example shows a *fractional factorial design* where only a subset of full choice sets (i.e., 12 choice tasks) are involved in the DCE, each with two alternatives. The alternative was described by *five attributes* (e.g., represented as A1, A2 to A5) and different attributes have various levels ranging from 2 (i.e., 0, 1) to 4 (i.e., 0, 1, 2, 4). By changing the levels of the attributes across different alternatives and the choice tasks, the DCE allows researchers to determine what influence the preference of individuals and the relative importance of various attributes (Street et al., 2005).

The very critical first step in conducting a DCE is constructing an *efficient* experimental design. There are two commonly used types of DCE designs: *full factorial design* and *fractional factorial design* (Mangham et al., 2009). A full factorial design

L. Shang and Y. Chandra, *Discrete Choice Experiments Using R*,
https://doi.org/10.1007/978-981-99-4562-7_7

Table 7.1 An example of an experimental design in DCE

Choice task #	Alternative 1					Alternative 2				
	A1	A2	A3	A4	A5	A1	A2	A3	A4	A5
1	1	0	0	1	2	0	1	1	1	3
2	1	0	0	2	0	2	1	1	1	2
3	2	1	1	0	3	0	0	0	0	2
4	2	0	0	0	1	2	1	0	0	1
5	0	0	1	0	3	2	1	1	1	0
6	2	0	0	2	3	1	0	1	1	2
7	2	1	1	1	0	2	0	1	1	1
8	0	1	1	1	3	1	1	1	2	3
9	0	0	1	2	3	0	1	0	0	0
10	1	0	1	1	0	1	0	0	2	3
11	0	0	1	1	2	1	1	1	0	3
12	1	1	1	2	0	0	1	1	0	2

can be created by including *all possible combinations of the levels for all attributes*. If we have k attributes, each with two levels, then the total number of possible alternative combinations is 2^k. In the demonstrative project that we illustrated in Chap. 6 (see Table 6.4), there is a total of *6 attributes*, 5 of them have two levels, and one of them has three levels. If we adopt a full factorial design, the total number of alternatives for this DCE design will be 96 ($2^5 \times 3^1 = 96$).

A full factorial design allows researchers to estimate both *main effects*, which refers to the direct independent effect of each individual attribute on the respondents' choices (i.e., dependent variable) (e.g., how different price levels influence individual preferences for alternative meat products), and *interaction effect*—that is the extent to which the variations influence respondents' behavior (i.e., dependent variable) in the combination of two or more attributes (e.g., the difference in nutrition level combined with the price difference).

However, in most cases, a full factorial design is *not practical* as it may be considered *too tedious* for respondents to evaluate too many combinations. Using our demonstrative project as an example, a full combination of all possibilities would result in a total of 96 profiles (i.e., $2 \times 2 \times 2 \times 2 \times 2 \times 3 = 96$). It becomes impossible to evaluate 96 profiles (or alternatives). Thus, a *fractional factorial design* is often used by selecting some or a subset from all possible alternatives from the full factorial design. A fractional factorial design is able to reduce the number of choice sets presented. However, researchers need to be aware that although the fractional factorial design is considered a more practical choice, such design only supports identifying the *main effects* and *some interactions* but *not all interactions* (Johnson et al., 2013) among attributes.

The quality of a fractional factorial design is often measured by relative design efficiency, a function of the variances and covariances of the parameter estimates

(Vanniyasingam et al., 2016). While presenting all possible combinations of attributes and their levels will result in 100% statistical efficiency, this is not practically feasible. We will need to utilize a fractional factorial design in many cases. For determining an efficient experiment fractional factorial design, researchers often use a measure known as *D-efficiency* or *D-optimality*—a measure of to what extent we will be able to estimate the effects of interest after implementing the experiment (Carlsson & Martinsson, 2003; Burgess & Street, 2005). This design is under the assumption that all options are equally attractive (Bush, 2014). The D-efficiency criterion allows for minimizing the variance–covariance matrix and maximizing the determinant of information matrix parameter estimation (Jaynes et al., 2016). Notably, this criterion requires researchers to make plausible assumptions on the relative magnitude and signs of the utility coefficients when these are expected to be non-zeroes (Yao et al., 2015). These assumptions may come either from a pilot survey, from prior studies, or from experts' opinions.

This measure can be computed automatically by a variety of R packages, for example, the package **idefix** and the package **AlgDesign**.

There are two ways to report D-efficiency in *absolute* or *relative* forms. Absolute D-efficiency refers to the D-score, which is a number to measure D-efficiency for a given experimental design, while a relative D-score refers to the comparison of multiple designs within a class. Most extant experimental design software (e.g., Ngene, JMP, Lighthouse Studio) utilizes algorithms to find a D-efficient design that contains the smallest possible design that identifies all the necessary parameters. The ideal D-score is 1, indicating a statistically efficient design, although anything above 0.8 is considered reasonable. An experimental design has a D-score of 1 when it is *balanced* and *orthogonal,* and we will discuss these in more detail below.

7.2 Balanced and Orthogonal Designs

A perfectly efficient design (with a D-score of 1) is *balanced*, meaning each attribute level *appears equally often* within an attribute across all choice tasks. A balanced design *minimizes the variance* in the parameter estimates (Kuhfeld & Tobias, 2005). To ensure balance, the total number of alternatives in the experimental design should be *evenly divisible* by the number of levels for each attribute. For example, in the example matrix shown in Table 7.1, as we have attributes with 2, 3, and 4 levels, attribute level balance could only be guaranteed if the design size (i.e., the number of choice tasks) is divisible by three and four, such as 12, 24, 36, etc. The attribute A1 has balanced levels as each level (i.e., 0, 1, 2) appears four times across the 12 choice tasks. According to Bliemer et al. (2019), attribute level balance is not strictly required, but it is often considered desirable to get good coverage over the data space.

An experimental design is considered *orthogonal* if *each pair of attribute levels appear equally often* across all pairs of attributes within the experimental design (Johnson et al., 2013). A full factorial design is considered *strictly* orthogonal as it consists of all possible combinations of alternatives. Orthogonality is a desired characteristic of experimental designs that demands strictly independent variation of levels across attributes. If all attributes have the same number of levels (e.g., all attributes have two levels), then the design matrix is called a *fixed-level* orthogonal array (Hedayat et al., 1999), otherwise known as *mixed* orthogonal arrays (Bliemer et al., 2009). Moreover, orthogonal arrays allow *blocking* of the design matrix as it could help maintain attribute level balance within each block. We will discuss *blocked design* in the next section.

In addition to balance and orthogonality, an efficient DCE design should also aim to have minimal overlap (Maddala et al., 2003; Mangham et al., 2009)—that is, to minimize the probability that an attribute level repeats itself in each choice set. Although overlap can simplify the construction of choice sets, it comes at the cost of reduced design efficiency. This is because it limits the amount of trade-off information obtained by the design (Johnson et al., 2013).

7.3 Blocked Fractional Factorial Designs

It is often the case that an experimental design contains more choice questions than a researcher wishes to ask each respondent. For a situation like this, the researcher may need to consider *blocking* the experimental design by *splitting the choice sets into different blocks*, which are partitions (usually equally sized) of the experimental design that contain only a limited number of choice sets for each respondent (Johnson et al., 2013). This is called a *heterogenous* design, where respondents receive different versions of questionnaires. While each respondent is subject to the same choice sets, we call this a *homogenous* design. For example, the study by Janssen et al. (2018) utilized a blocked fractional factorial design by separating 48 choice tasks into three 16-task survey versions, and each respondent will only be assigned to answer one version (i.e., one of the blocks). Their survey versions were selected to minimize the average collection between the version and attribute levels.

Blocking can promote response efficiency by reducing the burden and cognitive effort of each respondent who participates in a DCE experiment. Typically, an experiment design asks respondents to consider up to 18 choice sets, as 18 represents a practical limit of how many comparisons can be finished before respondents get bored and annoyed (Hanson et al., 2005). According to a review by de Bekker-Grob et al. (2012), most DCE studies in health economics had a mean number of choice sets per respondent of 14.

7.4 Labeled and Unlabeled DCE Designs

There are two different approaches to presenting DCE choice sets, either in a *labeled* form or *unlabeled* form. The labeled form involves assigning labels or titles that contain certain information regarding the alternative (de Bekker-Grob et al., 2012).

In a labeled design, the choice alternatives are labeled with attribute levels. Usually, *labels* tend to consist of brand names, region names, logos, product names, specific treatment or test names (in health economics), and types of technology, which function as heuristic cues (Hensher et al., 2005; Van Rijnsoever et al., 2015). A key advantage of using a labeled format is that alternatives in a DCE will be perceived as more realistic and better reflect real-world scenarios by respondents, adding validity to the results (de Bekker-Grob et al., 2012). However, sometimes, the label may trigger feelings and thoughts that do not necessarily match the observed attributes of the product or service under examination and influence an individual's decision-making (Van Rijnsoever et al., 2015). For example, in a study of consumer preferences for coffee, a label such as "organic" or "fair trade" might influence respondents' choices even if the taste and price of the coffee are held constant. This could be due to respondents' beliefs or values about these labels and their associations with particular attributes or qualities.

Using our demonstrative project that examines individuals' preferences for alternative meat products as an example, a labeled DCE may present choice sets in the following format (see Table 7.2). The "OmniPork" and "BeyondBeef" are what we meant by a labeled form in DCE design.

On the other hand, an unlabeled form involves assigning unlabeled alternatives in each choice set, for example, "Alternative Meat A", "Alternative Meat B", and "Alternative Meat C" (instead of "OmniPork" or "BeyondBeef"), which do not provide additional information on the alternatives. This type of design uses *generic titles* for the alternatives without convening any information to respondents. Such an unlabeled DCE design might be more robust as the alternatives may be less correlated with the attributes as those in labeled DCEs (because the names "Alternative A" and "Alternative B" do not provide any extra information to respondents). Consequently, respondents will decide by trading off attribute levels with minimum influence from the labels (de Bekker-Grob et al., 2012). Using our sample project, an unlabeled DCE may present choice sets as follows (see Table 7.3).

Table 7.2 A labeled choice set in DCE design

Attributes	OmniPork	Beyond beef
Additives	Without additives	With additives
Nutrition	25 g protein	40 g protein
Format	Raw	Ready-to-Eat
Vegan Label	With vegan label	No vegan label
Carbon Footprint	No info on how the product reduces greenhouse gas emission	20% reduction in greenhouse gas emission
Price	HK$ 25	HK$ 45
Which product do you prefer to buy?		
Product A ☐	Product B ☐	Neither of Them ☐

Table 7.3 An unlabeled choice set in DCE design

Attributes	Alternative meat product A	Alternative meat product B
Additives	Without additives	With additives
Nutrition	25 g protein	40 g protein
Format	Raw	Ready-to-Eat
Vegan Label	With vegan label	No vegan label
Carbon Footprint	No info on how the product reduces greenhouse gas emission	20% reduction in greenhouse gas emission
Price	HK$ 25	HK$ 45
Which product do you prefer to buy?		
Product A ☐	Product B ☐	Neither of Them ☐

A labeled DCE study can involve label effects into account and may be more suitable to predict as it is closer to real-world situations (Kruijshaar et al., 2009). However, unlabeled DCE also has its own merits. For instance, the design of an unlabeled DCE can be much *smaller* than a labeled DCE. This means that if each of the labeled options has a number of A attributes, each with L levels, and the choice sets are of size M, then there will be a total of L^{MA} possible choice sets. While if the options are not labeled, there will be only L^{A} possible choice sets. de Bekker-Grob and colleagues (2012) suggested that using a labeled form will significantly influence individual choices and reduce respondents' attention to the attributes. Unlabeled DCEs are considered more suitable and feasible to examine trade-offs between attributes of respondents who have little knowledge of the alternative labels, while a labeled form is more appropriate for explaining real-life choices.

7.5 Designing a DCE Using R Packages

After introducing important concepts and criteria involved in the experimental design of a DCE study, it is time to look into how we can utilize R packages to assist in the design process. There are various packages available in the R platform that supports the design and analysis of DCEs. In this book, we will introduce four popular packages which enable researchers to generate optimal designs for DCEs, which include (1) **support.CEs**, (2) **idefix**, (3) **choiceDes**, and (4) **DCEtool**. We will provide step-by-step instructions on how to utilize these packages for designing and constructing a DCE, starting with installing packages.

7.5.1 *Using the Package 'support.CEs' for DCE Design*

The package **support.CEs**[1] is developed by Aizaki (2012), and it provides seven basic functions for choice experiments in R, including:

- two functions for creating an experimental design based on orthogonal arrays;
- a function that converts an experimental design into a questionnaire format;
- a function for converting a design into a matrix;
- a function for organizing the dataset into a format that is suitable for implementing a conditional logic model;
- a function for calculating the goodness of fit measures of an estimated model;
- a function for calculating the marginal willingness to pay.

In this chapter, we will focus on the two functions that are related to experimental designs—***rotation.design*** and ***Lma.design***.

The function ***rotation.design()*** executes two methods, including (1) rotation design method, which uses the orthogonal main-effect array as the first alternative in each choice set and adds a constant to each attribute level of the first alternative to create one or more additional alternatives, and (2) a mix-and-match design method, which introduces a randomizing process in the rotation method that allows for more flexible and efficient use of the available choice sets rather than using a fixed set of choice sets for all respondents (Aizaki, 2012).

This function generates an unlabeled type of DCE design that contains generic attributes. On the other hand, the function ***Lma.design()*** directly creates a choice experiment design from an orthogonal main-effect array (Johnson et al., 2007), which generates a labeled type choice experiment design that contains both generic attributes and alternative-specific attributes.

First, we will introduce how to install the package **support.CEs** in R. As an R package, **support.CEs** can be easily installed in R base or R Studio if users are familiar with the basics of the R programming language. This also applies to other R packages that we will discuss later (i.e., **idefix**, **DCEtool**). For installing the package on the R platform (or alternatively on RStudio), users can type the following code in the R Console:

install.packages("support.CEs")

After typing the code in the console window, press *enter*, and the download and installation of R packages will start automatically. Depending on the user's internet speed and computer performance, this may take a few moments. It is also possible to install the package using the graphical user interface in R Studio by selecting the "Packages" tab in the lower right-hand corner of the screen, clicking on "Install," and

[1] For more information on the R package, please refer to the CRAN at https://cran.r-project.org/web/packages/support.CEs/index.html.

typing "**support.CEs**" in the search bar. Figure 7.1 shows a screenshot of executing the package installation code in R Studio.

After successfully installing the R package in your computer, the next step is to *run the package* by typing and executing the following code in the console window (this also applies to other R packages):

library(support.CEs)

The ***library()*** is a function in R that loads the names of R packages available inside an R workspace. Only those successfully installed R packages can be loaded. After launching the package **support.CEs**, we may then start designing our DCE with the assistance of the R package. We will first demonstrate an example of an *unlabeled* DCE design created by the rotation design method using our demonstrative project of alternative meat products. In this demonstrative project, we have a total of *six attributes*. Five of them have two attribute levels, and one of them has three attribute levels (see Table 7.4).

In this example, we wish to present our respondents with two alternative products plus the option "*Neither of them*" (Table 7.2), and thus we will set the number of alternatives as 2 in the code. To design an unlabeled DCE, we will utilize the function

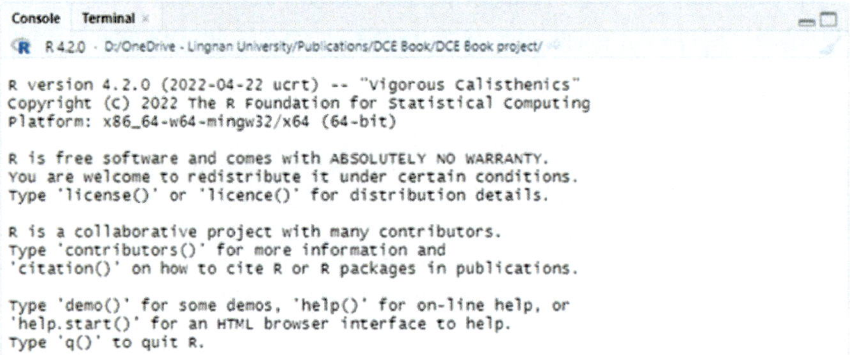

Fig. 7.1 A screenshot of installing the package support.CEs

Table 7.4 Attributes and attribute levels in the demonstrative project

Attributes	*Attribute levels (a 200-g of alternative meat)*
Additives	(1) Without additives
	(2) With additives
Nutrition	(1) 25 g of proteins
	(2) 40 g of proteins
Format	(1) Raw
	(2) Ready-to-eat
Vegan Label	(1) Without vegan label
	(2) With vegan label
Carbon Footprint	(1) No information on reduction in greenhouse emission
	(2) 20% of reduction in greenhouse gas emission
Price	(1) US$3.5
	(2) US$4.5
	(3) US$6.0

rotation.design() with the arguments assigned to the abovementioned conditions by typing and executing codes as follows:

```
unlabeled <- rotation.design(
  attribute.names = list(Additives = c("Without","With"),
                         Nutrition = c("25g","40g"),
                         Format = c("Raw","Ready"),
                         Carbon = c("No","20%"),
                         Vegan = c("No","Yes"),
                         Price = c("3.5","4.5","6.0")),
  nalternatives = 2,
  nblocks = 1,
  row.renames = FALSE,
  seed = 987, randomize = FALSE)
```

If you are using R Studio, we could enter the code in the Script Window first as it is easier to edit and save for future use (see Fig. 7.2).

The argument *attribute.names* specifies the names and levels of each attribute. In this case, we have six attributes, with levels specified in a list format. For example, the first attribute is called "Additives" and has two levels ("Without" and "With"), and the second attribute is called "Nutrition" and has two levels ("25 g" and "40 g"). Secondly, the argument *nalternatives* specifies the number of alternatives (or profiles) to generate per choice set. In this case, we have two alternatives per choice set. *nblocks* specifies the number of blocks (or choice sets) to generate. In this case, we have one

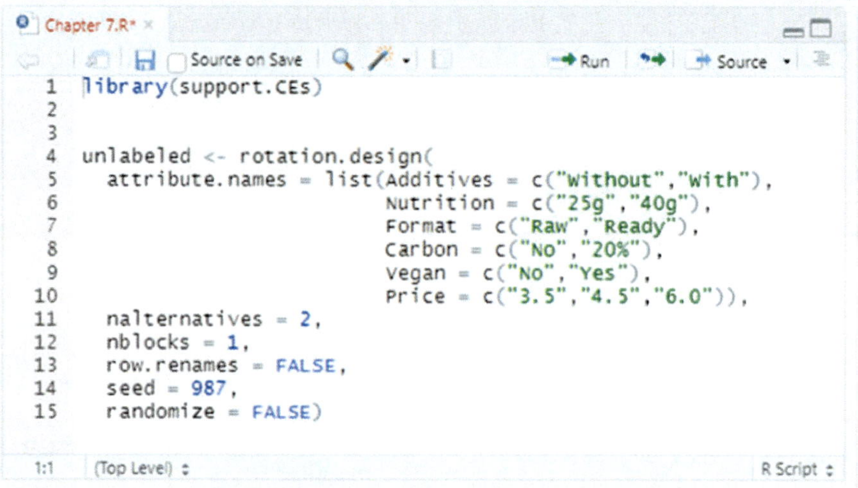

```
library(support.CEs)


unlabeled <- rotation.design(
  attribute.names = list(Additives = c("Without","With"),
                         Nutrition = c("25g","40g"),
                         Format = c("Raw","Ready"),
                         Carbon = c("No","20%"),
                         Vegan = c("No","Yes"),
                         Price = c("3.5","4.5","6.0")),
  nalternatives = 2,
  nblocks = 1,
  row.renames = FALSE,
  seed = 987,
  randomize = FALSE)
```

Fig. 7.2 Typing a code in the R script window (for R Studio)

block. *row.renames* specifies whether to rename the rows of the resulting design matrix. In this case, row names are not being renamed. *seed* specifies the random seed to use for generating the design. This ensures that the design is reproducible. Finally, *randomize* specifies whether to randomize the order of the generated profiles. In this case, the order is not randomized. Randomizing the design matrix can be useful to avoid potential order effects or other biases that may arise if the alternatives are always presented in the same order to respondents. Randomizing can help to ensure that the results are not affected by the order in which the alternatives are presented. However, it is important to note that randomization can also increase the sample size required for obtaining statistically significant results.

As shown in Fig. 7.2, the attributes of this demonstrative project are defined in the *attribute.names;* the number of alternatives is set to be 2, and the number of blocks is set to be 1, which means that *no blocking* is used to split the choice sets into different versions. After running the code, we could use the following command to print our design outputs:

unlabeled

Please note that *unlabeled* is the name that we gave to this DCE design for our demonstrative project, and you can change it to other names as you like. After running this code, the designed choice sets will be printed in the R console window, as shown in Fig. 7.3.

The function ***rotation.design()*** generates a total of 24 choice sets, each with two alternatives. The printed output indicates attribute levels involved in each alternative and thus can be utilized for constructing the DCE questionnaire (see Chap. 8 for more

```
Console  Terminal ×
R 4.2.0 · D:/OneDrive - Lingnan University/Publications/DCE Book/DCE Book project/
> unlabeled

Choice sets:
alternative 1 in each choice set
   BLOCK QES ALT Additives Nutrition Format Carbon Vegan Price
19     1   1   1      With       40g  Ready     No    No   6.0
17     1   2   1   Without       40g    Raw     No    No   3.5
13     1   3   1   Without       25g  Ready     No    No   4.5
11     1   4   1   Without       25g    Raw    20%    No   6.0
23     1   5   1      With       40g    Raw    20%    No   4.5
15     1   6   1   Without       40g  Ready    20%    No   6.0
6      1   7   1      With       25g  Ready    20%   Yes   3.5
14     1   8   1      With       40g  Ready     No   Yes   6.0
8      1   9   1      With       25g  Ready     No    No   3.5
12     1  10   1   Without       40g    Raw     No   Yes   6.0
7      1  11   1   Without       25g    Raw     No    No   3.5
1      1  12   1      With       25g    Raw    20%    No   4.5
22     1  13   1      With       40g    Raw    20%   Yes   4.5
4      1  14   1      With       25g    Raw     No   Yes   4.5
3      1  15   1      With       40g  Ready    20%   Yes   3.5
16     1  16   1   Without       25g  Ready     No   Yes   4.5
20     1  17   1      With       25g  Ready    20%    No   6.0
10     1  18   1   Without       25g    Raw    20%   Yes   6.0
24     1  19   1   Without       40g    Raw    20%   Yes   3.5
21     1  20   1   Without       25g  Ready    20%   Yes   3.5
5      1  21   1   Without       40g  Ready    20%    No   4.5
9      1  22   1      With       40g    Raw     No    No   3.5
18     1  23   1   Without       40g  Ready     No   Yes   4.5
2      1  24   1      With       25g    Raw     No   Yes   6.0

alternative 2 in each choice set
   BLOCK QES ALT Additives Nutrition Format Carbon Vegan Price
19     1   1   2   Without       25g    Raw    20%   Yes   3.5
17     1   2   2      With       25g  Ready    20%   Yes   4.5
13     1   3   2      With       40g    Raw    20%   Yes   6.0
11     1   4   2      With       40g  Ready     No   Yes   3.5
23     1   5   2   Without       25g  Ready     No   Yes   6.0
15     1   6   2      With       25g    Raw     No   Yes   3.5
6      1   7   2   Without       40g    Raw     No    No   4.5
14     1   8   2   Without       25g    Raw    20%    No   3.5
8      1   9   2   Without       40g    Raw    20%   Yes   4.5
12     1  10   2      With       25g  Ready    20%    No   3.5
7      1  11   2      With       40g  Ready    20%   Yes   4.5
1      1  12   2   Without       40g  Ready     No   Yes   6.0
22     1  13   2   Without       25g  Ready     No    No   6.0
4      1  14   2   Without       40g  Ready    20%    No   6.0
3      1  15   2   Without       25g    Raw     No    No   4.5
16     1  16   2      With       40g    Raw    20%    No   6.0
20     1  17   2   Without       40g    Raw     No   Yes   3.5
10     1  18   2      With       40g  Ready     No    No   3.5
24     1  19   2      With       25g  Ready     No    No   4.5
21     1  20   2      With       40g    Raw     No    No   4.5
5      1  21   2      With       25g    Raw     No   Yes   6.0
9      1  22   2   Without       25g  Ready    20%   Yes   4.5
18     1  23   2      With       25g    Raw    20%    No   6.0
2      1  24   2   Without       40g  Ready    20%    No   3.5
```

Fig. 7.3 Printed choice sets (Unlabeled DCE using support.CEs) details about questionnaire construction). The output suggests that each respondent will have to answer a total of 24 choice experiment questions, which might be considered to exceed the boredom threshold (Hanson et al., 2005). To *reduce* the number of choice sets for each respondent, **support.CEs** allows users to divide the choice sets into several blocks. For instance, if we want to divide the 24 choice sets generated in Fig. 7.3 into three blocks, we can simply change the value of *nblocks* in the syntax as follows:

```
unlabeled <- rotation.design(
  attribute.names = list(Additives = c("Without","With"),
                    Nutrition = c("25g","40g"),
                    Format = c("Raw","Ready"),
                    Carbon = c("No","20%"),
                    Vegan = c("No","Yes"),
                    Price = c("3.5","4.5","6.0")),
  nalternatives = 2,
  nblocks = 3,  ### this indicates dividing into 3 blocks
  row.renames = FALSE,
  seed = 987,  randomize = FALSE)
```

For using the mix-and-match method (with a randomization process that select a subset of choice sets for each respondent randomly to explore a large number of attribute combinations more efficiently) to generate an experimental design, we could change the value of FALSE after *randomize* into TRUE in the code:

```
unlabeled <- rotation.design(
  attribute.names = list(Additives = c("Without","With"),
                    Nutrition = c("25g","40g"),
                    Format = c("Raw","Ready"),
                    Carbon = c("No","20%"),
                    Vegan = c("No","Yes"),
                    Price = c("3.5","4.5","6.0")),
  nalternatives = 2,
  nblocks = 3,  ### this indicates dividing into 3 blocks
  row.renames = FALSE,
  seed = 987,  randomize = TRUE) ### changing to mix-
and-match method
```

In addition, to design a labeled DCE, we could use the function ***Lma.design()***, which is based on the L^{MA} method (L is the number of levels, M refers to the number of alternatives in each choice set, and A is the number of attributes for each alternative). This method creates the experimental design directly from a symmetric orthogonal main-effect array (Street et al., 2008).

Using our demonstrative project as an example, we treat the attribute *Additives* as an alternative-specific attributes whose attribute levels serve as labels of each alternative rather than one of the attributes. By doing this, we expect our respondents to select the most preferred option from two alternative meat products (*Product with Additives vs. Product without Additives*) and the option "Neither of them". Therefore, the argument **attribute.names** in the function ***Lma.design()*** is assigned by the other five attributes as follows:

```
labeled <- Lma.design(
  attribute.names = list(Nutrition = c("25g","40g"),
                         Format = c("Raw","Ready"),
                         Carbon = c("No","20%"),
                         Vegan = c("No","Yes"),
                         Price = c("3.5","4.5","6.0")),
  nalternatives = 2,
  nblocks = 1,
  row.renames = FALSE,
  seed = 987)
```

Similarly, we can print the output by running the syntax *labeled*. The printed output is shown in Fig. 7.4. Alternative *1* will be assigned a label of "**Product with Additives**" and *alternative* 2 will be given a label of "**Product without Additives**". We will need to manually replace the names of these alternatives when we construct the survey. This labeled DCE design contains a total of 36 choice sets, which is much larger than an unlabeled DCE design. Based on the printed choice sets, researchers can then convert them into a questionnaire format.

7.5.2 Using the Package 'idefix' for DCE Design

The second R package that we will introduce in this book is **idefix**,[2] which is a very new package developed by Traets et al. (2020). This package aims to generate efficient designs for DCEs based on the *multinomial logit model*, which is linked to the decision-making behavior via maximizing the utility (Li, 2011). According to the **idefix** developers, there are no other R packages that are suited to generate optimal designs for DCEs. The package **idefix** implements D-efficient to generate optimal designs. However, compared to the package **support.CEs**, **idefix** does not support the blocking design yet. Hence it cannot split the experiment design into smaller blocks.

[2] For more information on the R package, please refer to the CRAN at https://CRAN.R-project.org/package=idefix.

Fig. 7.4 Printed choice sets (Labeled DCE using support.CEs)

```
Choice sets:
alternative 1 in each choice set
   BLOCK QES ALT Nutrition Format Carbon Vegan Price
26     1   1   1       40g    Raw     No    No   4.5
19     1   2   1       40g  Ready    20%    No   4.5
35     1   3   1       40g    Raw    20%   Yes   3.5
17     1   4   1       40g  Ready     No   Yes   3.5
13     1   5   1       25g    Raw     No    No   4.5
30     1   6   1       25g    Raw     No   Yes   4.5
11     1   7   1       25g  Ready    20%   Yes   6.0
23     1   8   1       40g    Raw    20%   Yes   6.0
15     1   9   1       25g    Raw    20%   Yes   6.0
36     1  10   1       40g  Ready     No   Yes   4.5
6      1  11   1       25g  Ready    20%    No   3.5
14     1  12   1       40g  Ready    20%    No   3.5
33     1  13   1       25g    Raw    20%   Yes   4.5
29     1  14   1       25g    Raw     No   Yes   6.0
8      1  15   1       25g  Ready    20%   Yes   3.5
12     1  16   1       25g  Ready    20%   Yes   4.5
7      1  17   1       25g  Ready     No    No   4.5
31     1  18   1       25g    Raw     No   Yes   3.5
1      1  19   1       40g    Raw    20%   Yes   4.5
27     1  20   1       25g  Ready     No    No   3.5
28     1  21   1       25g  Ready    20%    No   4.5
22     1  22   1       25g    Raw    20%   Yes   3.5
4      1  23   1       40g    Raw    20%    No   4.5
3      1  24   1       40g  Ready     No   Yes   6.0
34     1  25   1       40g    Raw     No    No   6.0
16     1  26   1       40g    Raw     No    No   3.5
20     1  27   1       25g  Ready     No    No   6.0
10     1  28   1       25g    Raw     No    No   3.5
32     1  29   1       40g  Ready     No   Yes   6.0
24     1  30   1       40g  Ready    20%    No   6.0
21     1  31   1       40g  Ready     No   Yes   3.5
5      1  32   1       25g  Ready    20%    No   6.0
9      1  33   1       40g    Raw    20%    No   6.0
25     1  34   1       40g    Raw    20%    No   3.5
18     1  35   1       25g    Raw     No    No   6.0
2      1  36   1       40g  Ready     No   Yes   4.5

alternative 2 in each choice set
   BLOCK QES ALT Nutrition Format Carbon Vegan Price
26     1   1   2       40g  Ready    20%    No   4.5
19     1   2   2       40g  Ready     No   Yes   3.5
35     1   3   2       25g  Ready     No    No   6.0
17     1   4   2       40g    Raw    20%    No   6.0
13     1   5   2       25g    Raw     No    No   4.5
30     1   6   2       40g  Ready     No   Yes   4.5
11     1   7   2       25g  Ready    20%    No   4.5
23     1   8   2       25g  Ready     No    No   6.0
15     1   9   2       40g    Raw    20%   Yes   4.5
36     1  10   2       40g    Raw    20%    No   6.0
6      1  11   2       40g    Raw     No    No   6.0
14     1  12   2       40g  Ready     No   Yes   4.5
33     1  13   2       40g    Raw    20%   Yes   3.5
29     1  14   2       40g  Ready     No   Yes   6.0
8      1  15   2       25g  Ready    20%    No   3.5
12     1  16   2       25g  Ready    20%    No   3.5
7      1  17   2       25g  Ready    20%   Yes   6.0
31     1  18   2       40g  Ready     No   Yes   6.0
1      1  19   2       25g  Ready     No    No   4.5
27     1  20   2       25g  Ready    20%   Yes   4.5
28     1  21   2       40g    Raw     No    No   6.0
22     1  22   2       40g    Raw    20%   Yes   3.5
4      1  23   2       25g    Raw    20%   Yes   6.0
3      1  24   2       40g    Raw    20%    No   4.5
34     1  25   2       40g  Ready    20%    No   3.5
16     1  26   2       40g  Ready    20%    No   3.5
20     1  27   2       25g  Ready    20%   Yes   6.0
10     1  28   2       25g    Raw     No    No   3.5
32     1  29   2       25g    Raw     No   Yes   3.5
24     1  30   2       40g  Ready     No   Yes   3.5
21     1  31   2       25g    Raw     No   Yes   4.5
5      1  32   2       40g    Raw     No    No   4.5
9      1  33   2       25g    Raw    20%   Yes   6.0
25     1  34   2       25g    Raw    20%   Yes   4.5
18     1  35   2       25g    Raw     No    No   3.5
2      1  36   2       25g    Raw     No   Yes   3.5
```

To install this package, we could follow the same step illustrated above in 7.5.1. by typing and running the following code in the R console window:

```
install.packages("idefix")
```

To design an efficient DCE, we will need to use the function ***Profiles()*** in the package after specifying attributes and attribute levels we want to involve in the experimental design. Taking the alternative meat product in the demonstrative project for illustration, we can execute the following code for generating all possible combinations of attributes and levels as a starting point for designing an efficient DCE:

```
library(idefix)
at.lvls <- c(2,2,2,2,2,3)
c.type <- c("D","D","D","D","D", "C")
con.lvls <- list(c(3.5, 4.5, 6))
Profiles(lvls = at.lvls, coding = c.type, c.lvls = con.lvls)
```

In the above R code, the argument *at.lvls* specifies the number of levels for each attribute. In this case, we have six attributes, each with two levels (for the discrete attributes) or one continuous range (for the continuous attribute). *c.type* specifies the coding type for each attribute. "D" indicates that the attribute is a dummy variable (categorical), while "C" indicates that the attribute is continuous. *con.lvls* specifies the range of values for the continuous attribute. In this case, we have a single continuous attribute with levels 3.5, 4.5, and 6 (US$).

After running the code above, a total of 96 alternatives will be automatically printed and shown in the console window. The output is a matrix in which each row is a possible alternative (see Fig. 7.5 for a screenshot of partial combinations). With this matrix, users can then replace the numbers (i.e., 1, 0) with real attribute levels (i.e., 25 g proteins, 40 g proteins) for constructing the DCE questionnaire.

The function ***Profiles()*** in the package **idefix** has three arguments, which include (1) *lvls* that specifies how many attributes should be included, and how many levels each attribute will have, (2) *coding*, that is types of coding for each attribute, and (3) *c.lvls,* which is an optional argument that specifies the values of continuous attributes if any. For the argument *lvls*, we specified the number of attributes and their levels in *at.lvls* (e.g., a total of 5 attributes with two levels and one attribute with three levels). For the argument *coding*, categorical attributes with two or more levels can be coded as "D" (i.e., dummy coding) or "E" (i.e., effect coding that allows the assignment of different weights) and attributes that can be treated as continuous can be coded as "C". For example, the attribute Price in our demonstrative project is a continuous attribute with three levels. If our design consists of continuous attributes, we will need to utilize the argument *c.lvls* and specify the levels of those attributes as shown in the code above. If we drop the attribute Price, then we will no longer

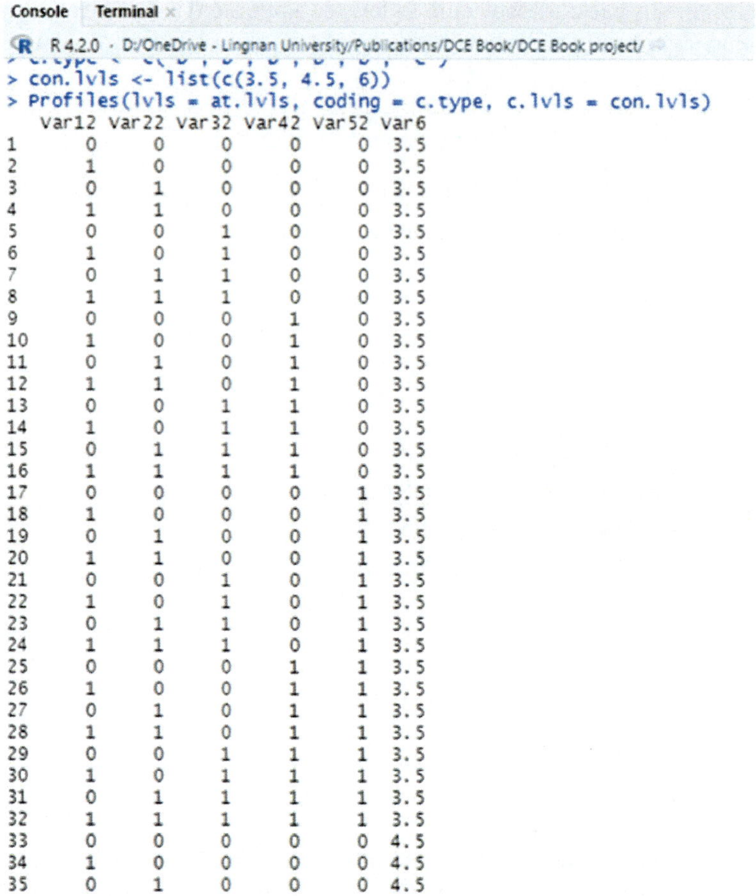

```
Console   Terminal
R 4.2.0 · D:/OneDrive - Lingnan University/Publications/DCE Book/DCE Book project/
> con.lvls <- list(c(3.5, 4.5, 6))
> Profiles(lvls = at.lvls, coding = c.type, c.lvls = con.lvls)
   var12 var22 var32 var42 var52 var6
1      0     0     0     0     0  3.5
2      1     0     0     0     0  3.5
3      0     1     0     0     0  3.5
4      1     1     0     0     0  3.5
5      0     0     1     0     0  3.5
6      1     0     1     0     0  3.5
7      0     1     1     0     0  3.5
8      1     1     1     0     0  3.5
9      0     0     0     1     0  3.5
10     1     0     0     1     0  3.5
11     0     1     0     1     0  3.5
12     1     1     0     1     0  3.5
13     0     0     1     1     0  3.5
14     1     0     1     1     0  3.5
15     0     1     1     1     0  3.5
16     1     1     1     1     0  3.5
17     0     0     0     0     1  3.5
18     1     0     0     0     1  3.5
19     0     1     0     0     1  3.5
20     1     1     0     0     1  3.5
21     0     0     1     0     1  3.5
22     1     0     1     0     1  3.5
23     0     1     1     0     1  3.5
24     1     1     1     0     1  3.5
25     0     0     0     1     1  3.5
26     1     0     0     1     1  3.5
27     0     1     0     1     1  3.5
28     1     1     0     1     1  3.5
29     0     0     1     1     1  3.5
30     1     0     1     1     1  3.5
31     0     1     1     1     1  3.5
32     1     1     1     1     1  3.5
33     0     0     0     0     0  4.5
34     1     0     0     0     0  4.5
35     0     1     0     0     0  4.5
```

Fig. 7.5 Printed possible combinations (using idefix)

have a continuous attribute in the experimental design. We can then modify the code into the following:

library(idefix)
at.lvls <- c(2,2,2,2,2)
c.type <- c("D","D","D","D","D")
Profiles(lvls = at.lvls, coding = c.type)

Printed output is shown in Fig. 7.6.

To generate an optimal experimental design, we could use the ***Modfed()*** function in the package **idefix**. According to Traets et al. (2020), this function has a total

```
Console   Terminal
R 4.2.0 · D:/OneDrive - Lingnan University/Publications/DCE Book/DCE Book project/
> Profiles(lvls = at.lvls, coding = c.type)
   Var12 Var22 Var32 Var42 Var52
1      0     0     0     0     0
2      1     0     0     0     0
3      0     1     0     0     0
4      1     1     0     0     0
5      0     0     1     0     0
6      1     0     1     0     0
7      0     1     1     0     0
8      1     1     1     0     0
9      0     0     0     1     0
10     1     0     0     1     0
11     0     1     0     1     0
12     1     1     0     1     0
13     0     0     1     1     0
14     1     0     1     1     0
15     0     1     1     1     0
16     1     1     1     1     0
17     0     0     0     0     1
18     1     0     0     0     1
19     0     1     0     0     1
20     1     1     0     0     1
21     0     0     1     0     1
22     1     0     1     0     1
23     0     1     1     0     1
24     1     1     1     0     1
25     0     0     0     1     1
26     1     0     0     1     1
27     0     1     0     1     1
28     1     1     0     1     1
29     0     0     1     1     1
30     1     0     1     1     1
31     0     1     1     1     1
32     1     1     1     1     1
```

Fig. 7.6 Printed possible combinations without continuous variables (using idefix)

of eleven arguments, of which seven arguments are optional. We listed the most commonly used functions as below:

- *cand.set*: a matrix that contains all possible alternatives that could be included in the experimental design;
- *n.sets*: the desired number of choice sets;
- *n.alts*: the number of alternatives in each choice set;
- *par.draws*: by specifying a numeric matrix in which each row is a draw from a multivariate parameter distribution, the DB-error will be optimized. DB-error quantifies how good or bad an experimental design is, and a design that minimizes DB-error is considered a D-optimal design (Hoyos, 2010)
- *alt.cte*: if alternative-specific constants (ASC), which refer to the "current situation" or status quo values (Boxall et al., 2009), the argument *alt.cte* should be specified by giving a binary vector that indicates whether an ASC should be present 1 or not 0 for each alternative.

- *no.choice*: if an opt-out option is desired in the experimental design, the *no.choice* argument can be set to **TRUE**. And the design will be optimized by taking into account that the last alternative of each choice set is an opt-out option.
- *best*: the default value of this argument is TRUE, which means that the output will only present the design with the lowest DB-error. If the value is set to be FALSE, then all optimized starting designs will be shown in the output.

Using our demonstrative alternative meat project (five attributes with two levels and one attribute with three levels) as an example, an optimal design with an opt-out option, "Neither of them", can be generated using the following code:

```
library(idefix)
at.lvls <- c(2,2,2,2,2,3)
c.type <- c("D","D","D","D","D", "C")
all <- Profiles(lvls = at.lvls, coding = c.type, c.lvls =
con.lvls)
con.lvls <- list(c(3.5, 4.5, 6))
set.seed(123)
mu <- c(-0.6, 1.2, -0.2, 0.6, 0.4, -0.4,-0.6)
sigma <- diag(length(mu))
M <- MASS::mvrnorm(n = 10, mu = mu, Sigma = sigma)
pd <- list(matrix (M[,1],ncol = 1), M[,2:7])
D <- Modfed(cand.set = all, n.sets=12, n.alts = 2,
no.choice = TRUE, alt.cte = c (0,1), par.draws = pd)
D ## print the design output
```

Please note that in the code above, we have specified parameters for all six attributes in *mu* before conducting the experiment, and *sigma* is a diagonal matrix of the same dimension as *mu* that specifies the variance of each random utility parameter. The ***mvrnorm()*** function from the **MASS** package is used to simulate random draws from this multivariate normal distribution, with *mu* and *sigma* specifying the mean and covariance structure of the distribution. These randomly generated coefficients are then used in the ***Modfed()*** function to generate a simulated DCE dataset with a specified number of choice sets and alternatives per choice set.

In particular, the first five parameters are for the first five attributes, respectively, and the last two parameters are for the last attribute, Price. For more information on specifying a prior distribution, users may refer to the paper by Traets et al. (2020), where the authors offered step-by-step instructions on how to specify an appropriate prior distribution, from gathering prior information through expert consultations and literature review to testing prior assumptions and robustness.

After running the code in the R console, an optimal design is printed, as shown in Fig. 7.7.

```
$design
            alt2.cte var12 var22 var32 var42 var52 var6
set1.alt1          0     1     1     0     0     0  3.5
no.choice          1     0     0     0     0     0  0.0
set2.alt1          0     1     0     1     0     0  3.5
no.choice          1     0     0     0     0     0  0.0
set3.alt1          0     0     1     0     1     0  3.5
no.choice          1     0     0     0     0     0  0.0
set4.alt1          0     0     0     0     0     0  3.5
no.choice          1     0     0     0     0     0  0.0
set5.alt1          0     1     0     1     0     0  6.0
no.choice          1     0     0     0     0     0  0.0
set6.alt1          0     0     1     0     0     1  3.5
no.choice          1     0     0     0     0     0  0.0
set7.alt1          0     0     1     0     0     1  6.0
no.choice          1     0     0     0     0     0  0.0
set8.alt1          0     0     1     1     0     0  3.5
no.choice          1     0     0     0     0     0  0.0
set9.alt1          0     1     0     1     1     0  3.5
no.choice          1     0     0     0     0     0  0.0
set10.alt1         0     0     0     0     1     1  3.5
no.choice          1     0     0     0     0     0  0.0
set11.alt1         0     1     0     1     0     1  3.5
no.choice          1     0     0     0     0     0  0.0
set12.alt1         0     1     0     0     0     0  3.5
no.choice          1     0     0     0     0     0  0.0

$error
[1] 7.701288

$inf.error
[1] 0

$probs
       Pr(alt1) Pr(no choice)
set1  0.5776077     0.4223923
set2  0.6854190     0.3145810
set3  0.5326709     0.4673291
set4  0.4071960     0.5928040
set5  0.5173150     0.4826850
set6  0.3706308     0.6293692
set7  0.3508375     0.6491625
set8  0.5569545     0.4430455
set9  0.7409559     0.2590441
set10 0.5210226     0.4789774
set11 0.6548105     0.3451895
set12 0.5386989     0.4613011
```

Fig. 7.7 Printed optimal design (using idefix)

Because the argument *best* is set as default (i.e., TRUE), the code above will only generate one optimal design with the lowest DB-error score. For generating all optimized starting designs, we could change the value to FALSE as in the code below:

```
library(idefix)
at.lvls <- c(2,2,2,2,2,3)
c.type <- c("D","D","D","D","D", "C")
all <- Profiles(lvls = at.lvls, coding = c.type, c.lvls =
con.lvls)
con.lvls <- list(c(3.5, 4.5, 6))
set.seed(123)
mu <- c(-0.6, 1.2, -0.2, 0.6, 0.4, -0.4,-0.6)
sigma <- diag(length(mu))
M <- MASS::mvrnorm(n = 10, mu = mu, Sigma = sigma)
pd <- list(matrix (M[,1],ncol = 1), M[,2:7])
D <- Modfed(cand.set = all, n.sets=12, n.alts = 2,
no.choice = TRUE, alt.cte = c (0,1), best = FALSE,
par.draws = pd)
D
```

7.5.3 Using the Package 'choiceDes' for DCE Design

Another package that supports the development of a DCE study is **chocieDes**, which is developed by Jack Horne (2018). This package extends the R package **AlgDesign,** which utilizes the Federov algorithm that improves an experimental design by tracking the amount of change in the determinant of the variance–covariance matrix in each step (Wheeler & Braun, 2019). Like **idefix**, the package **choiceDes** also provides functions for generating optimal designs. To install this package, we could type and run the following code in the R console window:

```
install.packages("choiceDes")
```

The core function of the package **choiceDes** is ***dcm.design()***, which generates an *optimal fractional factorial design* based on the vectors of factor lengths from a full factorial candidate set using the Federov algorithm. This function has seven arguments, of which four of them are important and need to be specified carefully:

- *cand*: a vector of factor lengths;
- *nb*: the number of blocks in the final experimental design;

- *sets*: the number of choice sets in each block;
- *alts*: the number of alternatives in each choice set;

To generate a D-optimal design using our demonstrative project, we could type and run the following code:

```
library(choiceDes)
levels <- c(2,2,2,2,2,3)
des <- dcm.design(levels,3,8,2)
des
```

In the code above, the number of blocks, the number of choice sets in each block, and the number of alternative in each choice is specified using the function ***dcm.design()***. In the code above, the resulting design will have 2 levels for each of the 5 attributes and 3 alternatives per choice set. The design will be divided into 3 blocks, each containing 8 choice sets.

After we execute the above code, an optimal design will be printed in the R console window with three sections, including a data frame consisting of the levels-coded design with blocks stacked in order, a list of effects-coded, blocked design, and diagnostics, and a list containing D-efficiency, the variance–covariance matrix, and parameter standard deviations from the effects-coded design. In most cases, we can look at the first and the third sections to have an optimal experimental design.

The screenshot below shows the printed result of the levels-coded design (See Fig. 7.8).

Figure 7.8 above presents an optimal DCE design that is divided into three blocks (i.e., *vers* in the printed output shows different blocks). In this design, each block has 8 choice sets (i.e., *card* in the printed output shows different choice sets. There is a total of 24 choice sets for three blocks). Each choice set has two alternatives (*tasks* refer to alternatives), and each alternative is described by six attributes (i.e., **X1**, **X2**, **X3** ... **X6**). The printed output also shows the D-efficiency score of the experimental design, as shown in Fig. 7.9:

Compared to other packages, the package **choiceDes** is more straightforward in terms of generating an optimal design using simple R codes (just several lines). But the package does not offer a function to convert the experimental design matrix into a questionnaire format, and the users need to manually replace the numbers in the matrix into real attribute levels for creating a DCE questionnaire. Moreover, as Traets et al. (2020) highlighted, the package does not take into account the dependency on the unknown preference parameters as the package **idefix**.

Fig. 7.8 Printed optimal design (using choiceDes)

```
> des
$levels
   card vers task x1 x2 x3 x4 x5 x6
45    1    1    1  2  1  1  2  2  3
18    1    1    1  1  2  1  1  1  2
8     2    1    2  2  2  2  2  1  1
6     2    1    2  2  2  1  2  1  1
23    3    1    3  1  2  2  2  1  2
16    3    1    3  1  2  2  2  2  1
1     4    1    4  1  1  1  1  1  1
40    4    1    4  1  2  2  2  1  3
29    5    1    5  1  1  1  2  2  2
25    5    1    5  1  1  1  1  2  2
42    6    1    6  2  1  2  1  2  3
26    6    1    6  2  1  1  1  2  2
3     7    1    7  2  1  2  1  1  1
36    7    1    7  2  2  2  1  1  3
11    8    1    8  2  1  2  1  2  1
46    8    1    8  1  2  1  2  2  3
31    9    2    1  2  1  2  2  2  2
47    9    2    1  2  2  1  2  2  3
17   10    2    2  2  1  1  1  1  2
19   10    2    2  2  2  1  1  1  2
38   11    2    3  2  1  1  2  1  3
43   11    2    3  1  2  2  1  2  3
12   12    2    4  1  2  2  1  2  1
32   12    2    4  1  2  2  2  2  2
7    13    2    5  1  1  2  2  1  1
10   13    2    5  2  2  1  1  2  1
13   14    2    6  2  1  1  2  2  1
34   14    2    6  1  2  1  1  1  3
37   15    2    7  1  1  1  2  1  3
4    15    2    7  2  2  2  1  1  1
48   16    2    8  1  1  2  2  2  3
20   16    2    8  1  1  2  1  1  2
33   17    3    1  1  1  1  1  1  3
27   17    3    1  1  1  2  1  2  2
28   18    3    2  2  2  2  1  2  2
24   18    3    2  2  2  2  2  1  2
21   19    3    3  1  2  1  2  1  2
41   19    3    3  2  2  1  1  2  3
35   20    3    4  1  1  2  1  1  3
15   20    3    4  1  1  2  2  2  1
5    21    3    5  2  1  1  2  1  1
14   21    3    5  1  2  1  2  2  1
30   22    3    6  2  2  1  2  2  2
9    22    3    6  1  1  1  1  2  1
39   23    3    7  2  1  2  2  1  3
44   23    3    7  2  2  2  1  2  3
22   24    3    8  2  1  2  2  1  2
2    24    3    8  1  2  1  1  1  1
```

7.5.4 Using the Package 'DCEtool' for DCE Design

The last R package that we will introduce is the package **DCEtool**, developed by Daniel Perez-Troncoso (2022). This is very different from the other three packages, as the experimental design, questionnaire construction, and data analysis of a DCE can be executed through a visual **user interface** (UI) in R that can execute R code on the backend rather than directly typing and running R codes in the console window. The UI of the package **DCEtool,** developed using the **shiny** package in R (Perez-Troncoso, 2022), is a standalone application on a webpage, and it provides users with a highly interactive and easy-to-use platform for creating a DCE design. This

```
$d.eff
$d.eff$D
[1] 0.02437356

$d.eff$v
           1          2          3          4          5           6           7
1 0.02083333 0.00000000 0.00000000 0.00000000 0.00000000  0.00000000  0.00000000
2 0.00000000 0.02083333 0.00000000 0.00000000 0.00000000  0.00000000  0.00000000
3 0.00000000 0.00000000 0.02083333 0.00000000 0.00000000  0.00000000  0.00000000
4 0.00000000 0.00000000 0.00000000 0.02083333 0.00000000  0.00000000  0.00000000
5 0.00000000 0.00000000 0.00000000 0.00000000 0.02083333  0.00000000  0.00000000
6 0.00000000 0.00000000 0.00000000 0.00000000 0.00000000  0.04166667 -0.02083333
7 0.00000000 0.00000000 0.00000000 0.00000000 0.00000000 -0.02083333  0.04166667

$d.eff$s
        1         2         3         4         5         6         7
0.1443376 0.1443376 0.1443376 0.1443376 0.1443376 0.2041241 0.2041241
```

Fig. 7.9 Printed D-efficiency scores (using choiceDes)

allows users who are not familiar with R programming or only know the basics also to have a chance to make use of R packages for designing and implementing DCEs.

To install the package, we could follow the steps introduced before by running the code below:

install.packages("DCEtool")

Before we can launch the package, we also need to install and activate the package Shiny:

install.packages("shiny")
library(shiny)

After successfully launching the package shiny, we can then launch the visual interface of DCEtool by executing the following syntax:

library(DCEtool)
DCEtool()

A DCEtool panel should then open in a separate window after we load the package (see Fig. 7.10).

This panel has a total of 7 sections in the panel menu, including *Home*, *Design Settings*, *Design matrix*, *Create a survey*, *Survey*, *Results*, and *About*. For generating a Bayesian D-efficient DCE design,[3] we can click the tab *Design settings*, inside which users can specify the attributes that are involved in their DCE study (see Fig. 7.11 for a screenshot of the Design settings).

[3] An efficient design with a specified prior distribution; see Kessels et al. (2011) for more information.

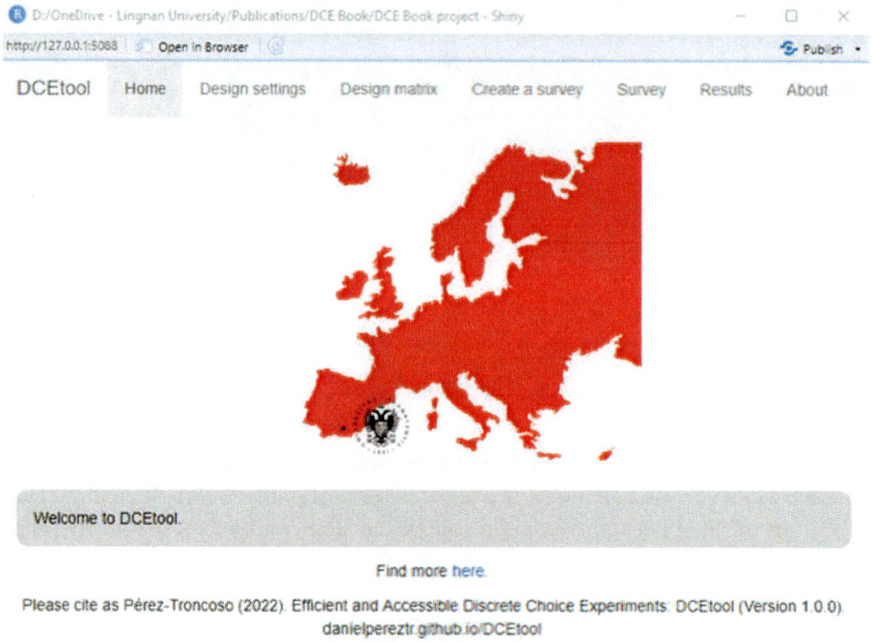

Fig. 7.10 The user interface of the package DCEtool

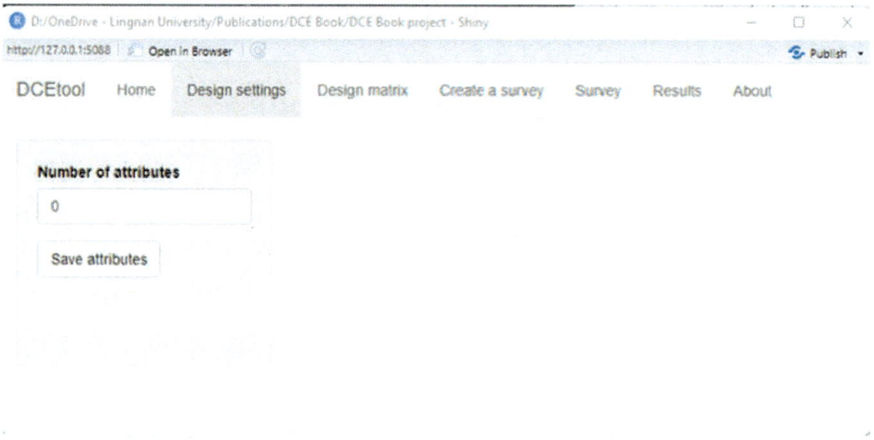

Fig. 7.11 A design settings tab in DCEtool package

Using our demonstrative project on alternative meat products as an example, we specified the number of attributes as six under the *Design settings* tab. And a new table will pop up that allows users to input the number of attribute levels for each attribute (see Fig. 7.12 for a screenshot).

After specifying attributes and corresponding levels, we can click on the button **Save attributes** to proceed to the next step of specifying the number of choice sets, the number of alternatives in each choice set, whether an opt-out option will be included, and whether to enable a Bayesian design or not (see Fig. 7.13 for a screenshot). For more information about the Bayesian approach, please refer to Kessels et al. (2011).

After we click on the button **Save settings**, we will be asked to provide more information about the prior coefficients (or parameters as in Sect. 7.5.2). The first five prior coefficients are for the first five attributes (each with two levels), respectively, and the last two are for the final attribute with three levels. Please note that the prior coefficients need to come from an existing DCE study (e.g., a pilot experiment or published research) with the same characteristics. Using prior coefficients allows for more efficient estimation of the choice model parameters and can help overcome issues with collinearity or data sparsity. Alternatively, we can use the default vector of zeros (Pérez-Troncoso, 2022) (as shown in Fig. 7.14).

After we finish specifying the experimental design, we could then click on the tab *Design matrix* in the menu and click on the button **Generate design** to start creating the design matrix (see Fig. 7.15).

A table with the experimental design matrix will be printed in the tab *Design matrix* (see Fig. 7.16).

In the design matrix, *task* refers to a choice set, and *alt* refers to alternatives involved in each choice set. As we include an opt-out option, each choice set has a total of three alternatives, while the third one is "Neither of them". Before saving the design and converting the design matrix into a questionnaire format, users can specify attribute names and level values. This will make questionnaire construction much easier. After specifying attribute names and levels, the printed design will be updated after clicking on the button **Change names in design matrix**. Please note that each attribute is missing one level (i.e., the first level you specified) in the columns to avoid multicollinearity (Perez-Troncoso, 2022). The printed result is shown in Fig. 7.17.

In general, the four packages we introduced in this chapter are all very useful for generating experimental DCE designs of DCEs. However, due to different features and functions, these packages have their own strengths as well as limitations. We summarized their key features in Table 7.5. Users can choose an R package that suits their DCE research needs the most.

Fig. 7.12 Specifying attributes and attribute levels in DCEtool

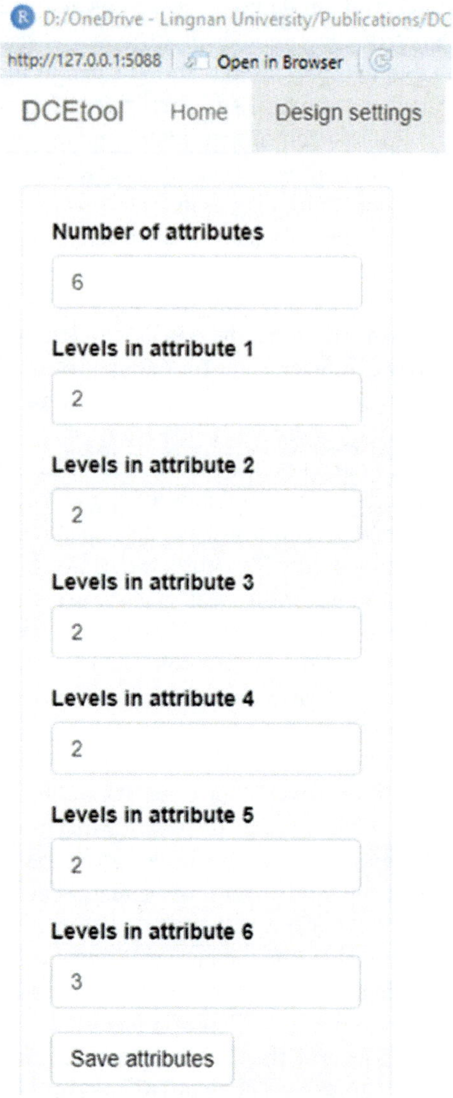

Fig. 7.13 Specifying choice sets using DCEtool package

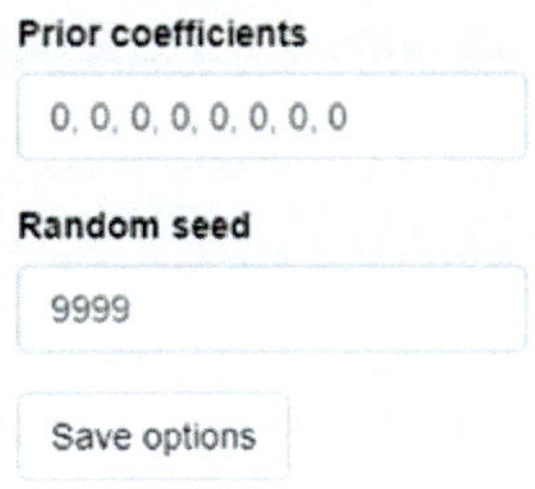

Fig. 7.14 Specifying prior coefficients using DCEtool package

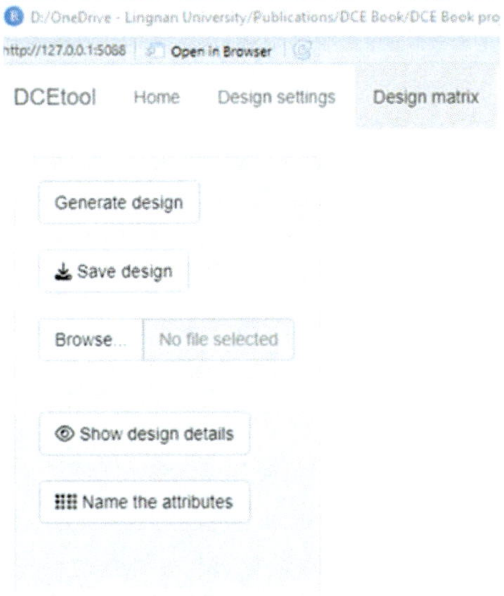

Fig. 7.15 Creating a design matrix using DCEtool package

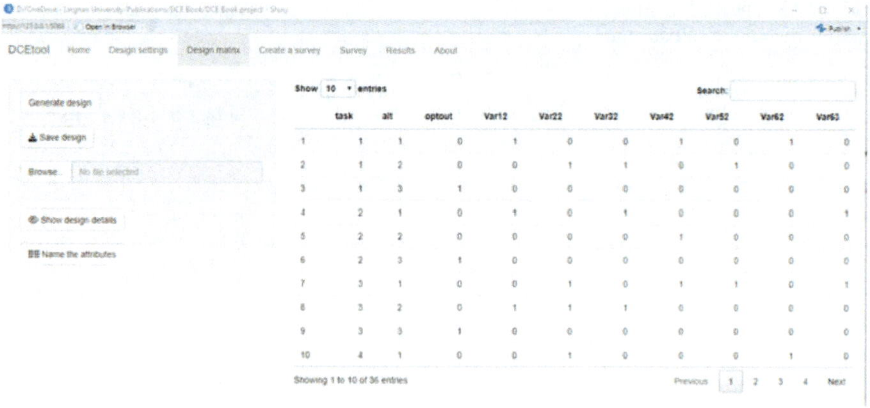

Fig. 7.16 Printed design matrix using DCEtool package

	task	alt	optout	With	40g	Ready	With	20%	4.5	6
1	1	1	0	1	0	0	1	0	1	0
2	1	2	0	0	1	1	0	1	0	0
3	1	3	1	0	0	0	0	0	0	0
4	2	1	0	1	0	1	0	0	0	1
5	2	2	0	0	0	0	1	0	0	0
6	2	3	1	0	0	0	0	0	0	0
7	3	1	0	0	1	0	1	1	0	1
8	3	2	0	1	1	1	0	0	0	0
9	3	3	1	0	0	0	0	0	0	0
10	4	1	0	0	1	0	0	0	1	0
11	4	2	0	0	0	0	0	0	0	0
12	4	3	1	0	0	0	0	0	0	0
13	5	1	0	0	1	1	1	0	1	0
14	5	2	0	1	0	0	0	1	0	0
15	5	3	1	0	0	0	0	0	0	0

Fig. 7.17 Printed design matrix with attribute names in DCEtool

Table 7.5 A summary of R packages for generating DCE designs

R packages	Developer(s)	Year of first release	Key features	CRAN
support.CEs	Hideo Aizaki	2012	It supports the generation of **orthogonal main-effect arrays** by creating an unlabelled DCE design using (1) a rotation design method and (2) a mix-and-match method. It can also create a labelled DCE design by realizing an LMA design method. It allows users to specify whether to use blocking design or not However, it does not support optimal experimental designs	https://cran.r-project.org/web/packages/support.CEs/index.html
idefix	Frits Traets & Daniel Gil	2017	It generates **efficient** designs for DCEs based on the multinomial logit model and the mixed logit model However, in order to produce efficient designs, users need to gather and specify an appropriate prior distribution and develop ideas about the sign and/ or magnitude of the model coefficients before designing the experiment. Moreover, the package does not support blocking design	https://cran.r-project.org/web/packages/idefix/index.html
choiceDes	Jack Horne	2014	This package is built upon AlgDes and generates **D-optimal designs** for linear models. It creates balanced and blocked experimental design by using a modified Fedorov algorithm However, it does not take into account the dependency on the unknown preference parameters, and it does not allow users to specify alternative specific constants ('opt-out' option or 'status quo' value)	https://cran.r-project.org/web/packages/choiceDes/index.html
DCEtool	Daniel Perez Troncoso	2021	This package has an interactive virtual user interface built on the shiny package. It is very user friendly to R beginners. It supports the generation of **efficient optimal and Bayesian DCEs.** It allows the users to specify prior coefficients to generate efficient designs However, it does not support the blocking design	https://cran.rstudio.com/web/packages/DCEtool/index.html

References

Aizaki, H. (2012). Basic functions for supporting an implementation of choice experiments in R. *Journal of Statistical Software, 50*, 1–24.

Bliemer, M. C., Rose, J. M., & Hensher, D. A. (2009). Efficient stated choice experiments for estimating nested logit models. *Transportation Research Part B: Methodological, 43*(1), 19–35.

Boxall, P., Adamowicz, W. L., & Moon, A. (2009). Complexity in choice experiments: Choice of the status quo alternative and implications for welfare measurement. *Australian Journal of Agricultural and Resource Economics, 53*(4), 503–519.

Bush, S. (2014). Optimal designs for stated choice experiments generated from fractional factorial designs. *Journal of Statistical Theory and Practice, 8*(2), 367–381.

Carlsson, F., & Martinsson, P. (2003). Design techniques for stated preference methods in health economics. *Health Economics, 12*(4), 281–294.

de Bekker-Grob, E. W., Ryan, M., & Gerard, K. (2012). Discrete choice experiments in health economics: A review of the literature. *Health Economics, 21*(2), 145–172.

Hanson, W. E., Creswell, J. W., Clark, V. L. P., Petska, K. S., & Creswell, J. D. (2005). Mixed methods research designs in counseling psychology. *Journal of Counseling Psychology, 52*(2), 224.

Hedayat, A. S., Sloane, N. J. A., & Stufken, J. (1999). *Orthogonal arrays: Theory and applications*. Springer Science & Business Media.

Hensher, D. A., Rose, J. M., Rose, J. M., & Greene, W. H. (2005). *Applied choice analysis: A primer*. Cambridge University Press.

Horne, J. (2018). choiceDes: Design functions for choice studies. R package version 0.9-3. Retrieved from https://CRAN.R-project.org/package=choiceDes

Hoyos, D. (2010). The state of the art of environmental valuation with discrete choice experiments. *Ecological Economics, 69*(8), 1595–1603.

Janssen, E. M., Hauber, A. B., & Bridges, J. F. (2018). Conducting a discrete-choice experiment study following recommendations for good research practices: An application for eliciting patient preferences for diabetes treatments. *Value in Health, 21*(1), 59–68.

Jaynes, J., Wong, W. K., & Xu, H. (2016). Using blocked fractional factorial designs to construct discrete choice experiments for healthcare studies. *Statistics in Medicine, 35*(15), 2543–2560.

Johnson, F., Kanninen, B., Bingham, M., & Özdemir, S. (2007). Experimental design for stated-choice studies. *Valuing Environmental Amenities using Stated Choice Studies*, 159–202.

Johnson, F. R., Lancsar, E., Marshall, D., Kilambi, V., Mühlbacher, A., Regier, D. A., ... & Bridges, J. F. (2013). Constructing experimental designs for discrete-choice experiments: Report of the ISPOR conjoint analysis experimental design good research practices task force. *Value in Health, 16*(1), 3–13.

Kessels, R., Jones, B., & Goos, P. (2011). Bayesian optimal designs for discrete choice experiments with partial profiles. *Journal of Choice Modelling, 4*(3), 52–74.

Kruijshaar, M. E., Essink-Bot, M. L., Donkers, B., Looman, C. W., Siersema, P. D., & Steyerberg, E. W. (2009). A labelled discrete choice experiment adds realism to the choices presented: Preferences for surveillance tests for Barrett esophagus. *BMC Medical Research Methodology, 9*(1), 1–10.

Kuhfeld, W. F., & Tobias, R. D. (2005). Large factorial designs for product engineering and marketing research applications. *Technometrics, 47*(2), 132–141.

Li, B. (2011). The multinomial logit model revisited: A semi-parametric approach in discrete choice analysis. *Transportation Research Part B: Methodological, 45*(3), 461–473.

Maddala, T., Phillips, K. A., & Reed Johnson, F. (2003). An experiment on simplifying conjoint analysis designs for measuring preferences. *Health Economics, 12*(12), 1035–1047.

Mangham, L. J., Hanson, K., & McPake, B. (2009). How to do (or not to do)... Designing a discrete choice experiment for application in a low-income country. *Health Policy and Planning, 24*(2), 151–158.

Perez-Troncoso. (2022). Design, conduct and analyze 'DCEs' from a virtual interface in shiny. Retrieved from https://github.com/danielpereztr/DCEtool

Pérez-Troncoso, D. (2022). Optimal sequential strategy to improve the precision of the estimators in a discrete choice experiment: A simulation study. *Journal of Choice Modelling, 43*, 100357.

Street, D. J., Burgess, L., & Louviere, J. J. (2005). Quick and easy choice sets: Constructing optimal and nearly optimal stated choice experiments. *International Journal of Research in Marketing, 22*(4), 459–470.

Street, D. J., Burgess, L., Viney, R., & Louviere, J. (2008). Designing discrete choice experiments for health care. In *Using discrete choice experiments to value health and health care* (pp. 47–72). Springer.

Traets, F., Sanchez, D. G., & Vandebroek, M. (2020). Generating optimal designs for discrete choice experiments in R: The idefix package. *Journal of Statistical Software, 96*, 1–41.

Van Rijnsoever, F. J., Van Mossel, A., & Broecks, K. P. (2015). Public acceptance of energy technologies: The effects of labeling, time, and heterogeneity in a discrete choice experiment. *Renewable and Sustainable Energy Reviews, 45*, 817–829.

Vanniyasingam, T., Cunningham, C. E., Foster, G., & Thabane, L. (2016). Simulation study to determine the impact of different design features on design efficiency in discrete choice experiments. *British Medical Journal Open, 6*(7), e011985.

Weber, S. (2021). A step-by-step procedure to implement discrete choice experiments in Qualtrics. *Social Science Computer Review, 39*(5), 903–921.

Wheeler, B., & Braun, M. J. (2019). Package 'AlgDesign.' *R Project for Statistical Computing, 1*, 1–25.

Yao, R. T., Scarpa, R., Rose, J. M., & Turner, J. A. (2015). Experimental design criteria and their behavioural efficiency: An evaluation in the field. *Environmental and Resource Economics, 62*, 433–455.

Chapter 8
Research Design and Data Collection in DCE

This chapter discusses and demonstrates the techniques to design and construct a DCE questionnaire and collect data using survey software such as Qualtrics. We will also discuss issues such as sampling frame, choices of sampling approach, minimum sample size, and questionnaire administration. This follows the best practices in generating a design matrix using various R packages discussed in Chap. 7.

8.1 Designing DCE Choice Sets in Questionnaire Format

After generating a design matrix[1] via R packages (see the step-by-step procedures in Chap. 7) (Weber, 2021), the next step is to convert the design matrix into a questionnaire format. In the questionnaire formatting process, researchers can manually construct choice sets one by one by replacing values (e.g., attribute levels with 0 and 1) in the design matrix with actual attribute names and levels. For instance, using an example of a printed optimal design[2] generated by R package **idefix** (see Fig. 8.1), a total of 12 choice sets were generated, each with two choice alternatives ("Alt1 and Alt2") that are described by six attributes (i.e., var12, var22, var 32, var42, var52, var6) and an "opt-out" option ("no.choice").

To convert the choice alternatives into a questionnaire format, we can manually replace the values of "0" and "1" with the real attribute levels (i.e., "25 g" and "40 g"; "raw" and "ready-to-eat").

[1] A table that specifies the attribute levels and profiles used in the choice sets presented to participants.

[2] An experimental design that aims to maximize the efficiency of parameter estimation with the smallest possible sample size or the fewest number of choice sets per respondent. A fractional factorial design with high D-efficiency scores can be considered optimal.

L. Shang and Y. Chandra, *Discrete Choice Experiments Using R*,
https://doi.org/10.1007/978-981-99-4562-7_8

```
R 4.2.0 · D:/OneDrive - Lingnan University/Publications/DCE Book/DCE Book project/
$design
           alt3.cte var12 var22 var32 var42 var52 var6
set1.alt1         0     1     1     1     1     1  3.5
set1.alt2         0     1     0     1     0     0  3.5
no.choice         1     0     0     0     0     0  0.0
set2.alt1         0     1     0     0     1     0  3.5
set2.alt2         0     1     1     1     1     0  3.5
no.choice         1     0     0     0     0     0  0.0
set3.alt1         0     1     0     1     1     1  3.5
set3.alt2         0     1     0     0     0     0  3.5
no.choice         1     0     0     0     0     0  0.0
set4.alt1         0     0     0     1     1     0  3.5
set4.alt2         0     1     1     1     0     0  3.5
no.choice         1     0     0     0     0     0  0.0
set5.alt1         0     1     1     0     1     0  3.5
set5.alt2         0     0     0     1     1     0  3.5
no.choice         1     0     0     0     0     0  0.0
set6.alt1         0     1     0     1     0     1  3.5
set6.alt2         0     1     0     1     1     0  3.5
no.choice         1     0     0     0     0     0  0.0
set7.alt1         0     1     0     1     0     0  3.5
set7.alt2         0     1     1     1     1     0  6.0
no.choice         1     0     0     0     0     0  0.0
set8.alt1         0     1     0     1     1     0  4.5
set8.alt2         0     0     0     1     1     0  3.5
no.choice         1     0     0     0     0     0  0.0
set9.alt1         0     1     0     1     1     1  3.5
set9.alt2         0     1     1     1     0     0  4.5
no.choice         1     0     0     0     0     0  0.0
set10.alt1        0     1     0     1     1     0  4.5
set10.alt2        0     1     1     1     1     0  3.5
no.choice         1     0     0     0     0     0  0.0
set11.alt1        0     1     0     0     1     0  3.5
set11.alt2        0     0     1     1     0     1  3.5
no.choice         1     0     0     0     0     0  0.0
set12.alt1        0     1     0     1     1     0  3.5
set12.alt2        0     1     1     0     0     1  3.5
no.choice         1     0     0     0     0     0  0.0

$error
[1] 4.942789
```

Fig. 8.1 Printed optimal design (using idefix)

For instance, the first choice set (including *set1.alt1*, *set1.alt2*, and *no.choice*) shown in Fig. 8.1 can be converted into a questionnaire format as below (see Table 8.1), which will be presented to respondents to respond.

The construction of the questionnaire in Table 8.1 works in the following way. First, ignore the column "alt3.cte", and pay attention to the column and their actual attribute name which are "var12 = Additives, var22 = Nutrition, var32 = Format, var42 = Vegan Label, var52 = Carbon Footprint, and var6 = Price" sequentially. For the first alternative, "*set1.alt1*", which contains the values of "1" (meaning: Has Additives), "1" (meaning: 40 g proteins), "1" (meaning: Ready-to-go), "1" (meaning: Has vegan label), "1" (meaning: 20% reduction in greenhouse emission), and "3.5" (meaning: 3.5 USD) (see the first row), these will be presented in the questionnaire using the actual attribute level names and values (read in vertical order) which are

Table 8.1 A sample choice set in a questionnaire format

Attributes	Alternative meat product A	Alternative meat product B
Additives	Has additives	Has additives
Nutrition	40 grams proteins	25 grams proteins
Format	Ready-to-go	Ready-to-go
Vegan label	Has vegan label	No vegan label
Carbon footprint	20% reduction in greenhouse gas emission	No reduction in greenhouse gas emission
Price	3.5 USD	3.5 USD
Which product do you prefer to buy?		
Product A ☐	Product B ☐	Neither of them ☐

"Has Additives", "40 g proteins", "Ready-to-go", "Has vegan label", "20% reduction in greenhouse emission", and *"3.5 USD"* (they are presented horizontally in the choice set). This choice alternative needs a name for it to make sense to respondents and thus we label it as "*Alternative Meat Product A*." The same approach is applied to construct "Alternative Meat Product B" and its attribute levels: "Has Additives, 25 g proteins, Ready-to-go, No vegan label, No reduction in greenhouse gas emission, 3.5 USD."

However, manually replacing attribute names and attribute levels one by one can be very time and energy-consuming. Alternatively, some functions in the R packages, such as the package **support.CEs**, **idefix**, and **DCEtool** to assist researchers in converting the design matrix into a questionnaire format *automatically*.[3] The main difference between packages in converting design matrices into a questionnaire format lies in the level of customization and flexibility offered to the user. The **support.CEs** package provides functions for generating different types of questionnaires for labeled and unlabeled designs. It offers some customization options for formatting and labeling questions, but it has limitations in customizing the questionnaire's layout and structure.

On the other hand, **idefix** and **DCEtool** offer more flexibility and customization options for generating questionnaires. For example, in **idefix**, users can define the layout and structure of the questionnaire using HTML and LaTeX code and customize the appearance of the questionnaire by specifying fonts, colors, and other design elements. In **DCEtool**, users can customize the format and appearance of questions and response options, as well as the instructions and introductory text. The choice of package to use will depend on the specific needs and preferences of the researcher. In this book, the focus is mainly on the use of **support.CEs** for designing DCEs and analyzing DCE data. However, other packages will be briefly introduced as well.

[3] Some other packages that can be used to automate the converting process include **Mlogit** and **dplyr**.

8.1.1 Designing DCE Choice Sets Using Support.CEs Package

The R package **support.CEs** (Aizaki, 2012) is a more user-friendly package (than other packages) to help with questionnaire construction for DCE studies. It has a function called ***questionnaire()***, which converts a choice experiment design into choice experiment questions that can be directly used in a questionnaire survey. Using this package, researchers do not need to manually label the attribute labels and levels as shown in Table 8.1.

To use the function, we can simply type additional code (see the last line in the R code below) into the bottom line of the R console after we specify the experimental design using the package:

```
library(support.CEs)
unlabeled <- rotation.design(
  attribute.names = list(Additives = c("Without","With"),
                         Nutrition = c("25g","40g"),
                         Format = c("Raw","Ready"),
                         Carbon = c("No","20%"),
                         Vegan = c("No","Yes"),
                         Price = c("3.5","4.5","6.0")),
  nalternatives = 2,
  nblocks = 1,
  row.renames = FALSE,
  seed = 987,
  randomize = FALSE)

##new code for creating questionnaire
questionnaire(choice.experiment.design = unlabeled)
```

The code *nalternatives* = 2 means that we want to create two alternatives in each choice set (e.g., Alternative Meat Product A vs B). We set *nblocks* = *1* because we do not have too many choice alternatives that need to be divided into different blocks (e.g., if we do, we can set *nblocks* to 2 or 3 depending on the number of alternatives included in a study).

After running the code, the choice sets will be automatically printed in the R Console, with each alternative specified by actual attributes and attribute levels. Figure 8.2 shows a screenshot of the partial printed output.

As can be seen in Fig. 8.2, **support.CE** package will make researchers' lives easier when constructing a DCE questionnaire. This automated approach in constructing the DCE questionnaire is not only faster to do but also eliminates the potential errors when manually labeling and filling in the values for each choice alternative.

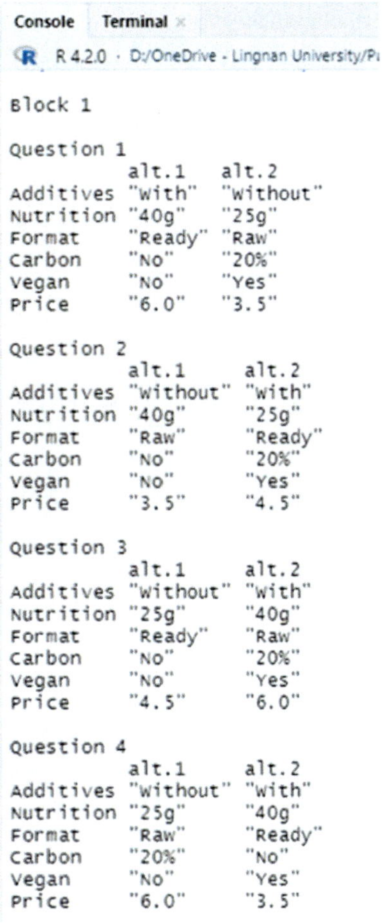

```
Console  Terminal
R 4.2.0 · D:/OneDrive - Lingnan University/P

Block 1

Question 1
           alt.1    alt.2
Additives "With"   "Without"
Nutrition "40g"    "25g"
Format    "Ready"  "Raw"
Carbon    "No"     "20%"
Vegan     "No"     "Yes"
Price     "6.0"    "3.5"

Question 2
           alt.1     alt.2
Additives "Without" "With"
Nutrition "40g"     "25g"
Format    "Raw"     "Ready"
Carbon    "No"      "20%"
Vegan     "No"      "Yes"
Price     "3.5"     "4.5"

Question 3
           alt.1     alt.2
Additives "Without" "With"
Nutrition "25g"     "40g"
Format    "Ready"   "Raw"
Carbon    "No"      "20%"
Vegan     "No"      "Yes"
Price     "4.5"     "6.0"

Question 4
           alt.1     alt.2
Additives "Without" "With"
Nutrition "25g"     "40g"
Format    "Raw"     "Ready"
Carbon    "20%"     "No"
Vegan     "No"      "Yes"
Price     "6.0"     "3.5"
```

Fig. 8.2 Printed choice sets using support.CEs package

8.1.2 Designing DCE Choice Sets Using Idefix Package

Similarly, the package **idefix** (Traets et al., 2020) has a ***Decode()*** function, which helps researchers transform a coded design matrix generated with the *Modfed* function into a design with specified attributes and attribute levels, ready to be used in a questionnaire format. To use the function, researchers need to specify each attribute and its levels in their DCEs.

For instance, using the alternative meat demonstrative project as an example, the first attribute is "Additives" with two different values: without additives and with additives (coded as the first "2" in the *at.lvls*). The second attribute represents "Nutrition" which consists of two values: 25 grams and 40 grams proteins (coded as the second "2" in the *at.lvls*). The third attribute is the "Format" of alternative meat products, which can be either raw or ready-to-eat (coded as the third "2" in the *at.lvls*). The next attribute "Vegan" also has two values, either has a vegan label on the product's packaging or does not have a vegan label (coded as the fourth "2" in the *at.lvls*). The fifth attribute is "Carbon Footprint", which also has two different levels: either a 20% reduction in carbon emissions or no information about the reduction in carbon emissions (coded as the fifth "2" in the *at.lvls*). Finally, we have a price attribute that has three levels: \$3.5 USD, \$4.5 USD, and \$6 USD (coded as "3" in the *at.lvls*). An additional new code for generating choice sets in a readable format is added after the experimental design:

```
library(idefix)
at.lvls <- c(2,2,2,2,2,3)
c.type <- c("D","D","D","D","D", "C")
all <- Profiles(lvls = at.lvls, coding = c.type, c.lvls =
con.lvls)
con.lvls <- list(c(3.5, 4.5, 6))
set.seed(123)
mu <- c(-0.6, 1.2, -0.2, 0.6, 0.4, -0.4,-0.6)
sigma <- diag(length(mu))
M <- MASS::mvrnorm(n = 10, mu = mu, Sigma = sigma)
pd <- list(matrix (M[,1],ncol = 1), M[,2:7])
D <- Modfed(cand.set = all, n.sets=12, n.alts = 3,
no.choice = TRUE, alt.cte = c (0,0,1), par.draws = pd)
D

## new syntax for creating the questionnaire
design <- D$design
al <- list(
  c("WithoutAdd","WithAdd"),
  c("25g","40g"),
  c("raw","ready-to-eat"),
  c("NoVegan","Vegan"),
  c("zeroReduction","20%reduction"),
  c("3.5USD","4.5USD","6USD"))

Decode(des=design,n.alts = 3, lvl.names = al,
coding=c.type, c.lvls = con.lvls,
     alt.cte = c(0,0,1),no.choice = 3)
```

In the above R code, we use the ***Decode()*** function from the **idefix** package to convert the design matrix generated by the ***Modfex()*** function into a questionnaire format with the actual attribute levels. We created a new object *al* that specifies the names of the levels for each attribute and then called it in the argument *lvl.names* of the function ***Decode()***. After executing the codes above, the DCE experimental design will be printed on the R console in a readable format. This then allows researchers to easily convert it into a questionnaire format for data collection (see Fig. 8.3). This way of displaying the experimental design makes it more intuitive to the researchers to quickly know what each choice alternative consists of in terms of attribute levels.

```
$design
                    V1  V2           V3      V4            V5   V6
set1.alt1      WithAdd 40g ready-to-eat   Vegan  20%reduction  3.5
set1.alt2      WithAdd 25g ready-to-eat NoVegan zeroReduction  3.5
no.choice         <NA> <NA>        <NA>    <NA>          <NA> <NA>
set2.alt1      WithAdd 25g          raw   Vegan zeroReduction  3.5
set2.alt2      WithAdd 40g ready-to-eat   Vegan zeroReduction  3.5
no.choice.1       <NA> <NA>        <NA>    <NA>          <NA> <NA>
set3.alt1      WithAdd 25g ready-to-eat   Vegan  20%reduction  3.5
set3.alt2      WithAdd 25g          raw NoVegan zeroReduction  3.5
no.choice.2       <NA> <NA>        <NA>    <NA>          <NA> <NA>
set4.alt1   WithoutAdd 25g ready-to-eat   Vegan zeroReduction  3.5
set4.alt2      WithAdd 40g ready-to-eat NoVegan zeroReduction  3.5
no.choice.3       <NA> <NA>        <NA>    <NA>          <NA> <NA>
set5.alt1      WithAdd 40g          raw   Vegan zeroReduction  3.5
set5.alt2   WithoutAdd 25g ready-to-eat   Vegan zeroReduction  3.5
no.choice.4       <NA> <NA>        <NA>    <NA>          <NA> <NA>
set6.alt1      WithAdd 25g ready-to-eat NoVegan  20%reduction  3.5
set6.alt2      WithAdd 25g ready-to-eat   Vegan zeroReduction  3.5
no.choice.5       <NA> <NA>        <NA>    <NA>          <NA> <NA>
set7.alt1      WithAdd 25g ready-to-eat NoVegan zeroReduction  3.5
set7.alt2      WithAdd 40g ready-to-eat   Vegan zeroReduction    6
no.choice.6       <NA> <NA>        <NA>    <NA>          <NA> <NA>
set8.alt1      WithAdd 25g ready-to-eat   Vegan zeroReduction  4.5
set8.alt2   WithoutAdd 25g ready-to-eat   Vegan zeroReduction  3.5
no.choice.7       <NA> <NA>        <NA>    <NA>          <NA> <NA>
set9.alt1      WithAdd 25g ready-to-eat   Vegan  20%reduction  3.5
set9.alt2      WithAdd 40g ready-to-eat NoVegan zeroReduction  4.5
no.choice.8       <NA> <NA>        <NA>    <NA>          <NA> <NA>
set10.alt1     WithAdd 25g ready-to-eat   Vegan zeroReduction  4.5
set10.alt2     WithAdd 40g ready-to-eat   Vegan zeroReduction  3.5
no.choice.9       <NA> <NA>        <NA>    <NA>          <NA> <NA>
set11.alt1     WithAdd 25g          raw   Vegan zeroReduction  3.5
set11.alt2  WithoutAdd 40g ready-to-eat NoVegan  20%reduction  3.5
no.choice.10      <NA> <NA>        <NA>    <NA>          <NA> <NA>
set12.alt1     WithAdd 25g ready-to-eat   Vegan zeroReduction  3.5
set12.alt2     WithAdd 40g          raw NoVegan  20%reduction  3.5
no.choice.11      <NA> <NA>        <NA>    <NA>          <NA> <NA>
```

Fig. 8.3 Printed choice sets using idefix package

8.1.3 Designing DCE Choice Sets Using DCEtool Package

Alternatively, researchers can use the R package **DCEtool** to generate a DCE questionnaire from the interactive user interface (Perez-Troncoso, 2022). Once we specify the experimental design (i.e., number of attributes, attribute levels, alternatives per choice set, and number of total choice sets), we will have a design matrix with real attribute names and level values (see Fig. 7.16 in Chap. 7). After that, we can directly launch a local survey from the user interface by clicking on the tab "**Create a survey**". In the DCEtool panel, we can type the introductory text that we want to demonstrate to respondents at the beginning of the questionnaire and an end text as well. Moreover, we can label each alternative by giving them a name (see Fig. 8.4).

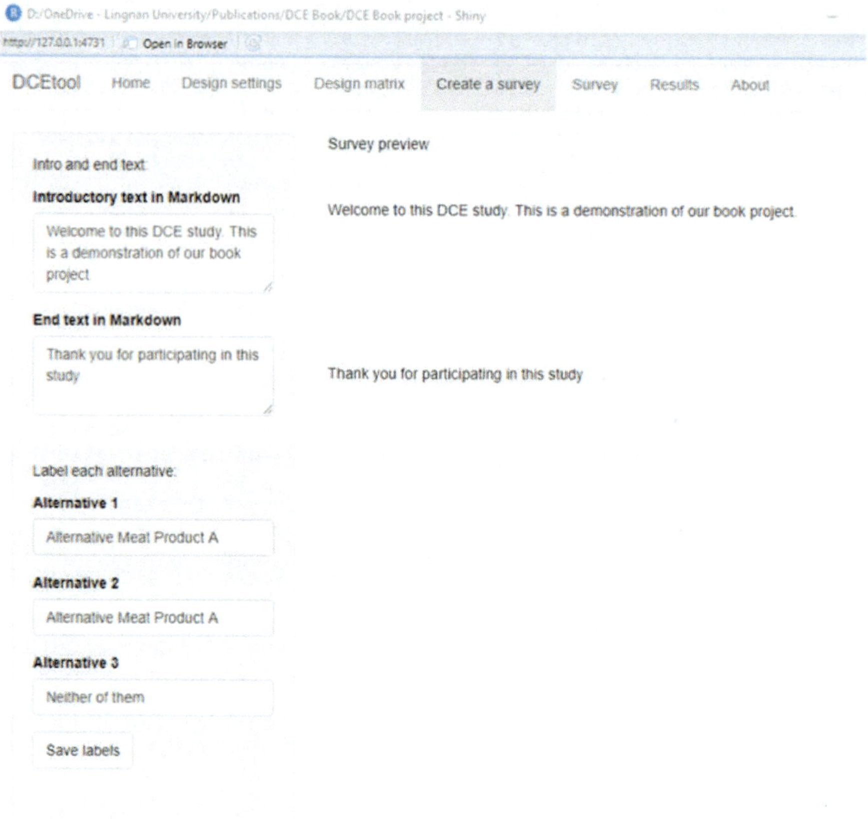

Fig. 8.4 Creating a survey in DCEtool package

After we click on the button, **Save labels**, a preview of the first choice set will appear, as shown in Fig. 8.5.

To launch a local survey, we can go to the next tab "**Survey**", and we will be asked to decide between three options[4] to indicate our preferred serial mode (see Fig. 8.6), which include (1) "No", a *usual* or *standard* DCE where the same choice sets are repeated for all respondents, (2) "Bliemer & Rose (each respondent)", a serial DCE in which a conditional logit will be estimated after each response and the coefficients will be used to adjust the next DCE (Bliemer & Rose, 2010), and (3) "Each 5 respondents", which means that a conditional logit will be estimated after every five new responses and the coefficients will be used to tune the next DCE (Pérez-Troncoso, 2022).

For our sample project, we will use the *usual or standard* DCE design by presenting the same choice sets to each respondent. After clicking on the button

---

[4] You can add more options/alternatives in the experimental design stage. See Chap. 7 for more information.

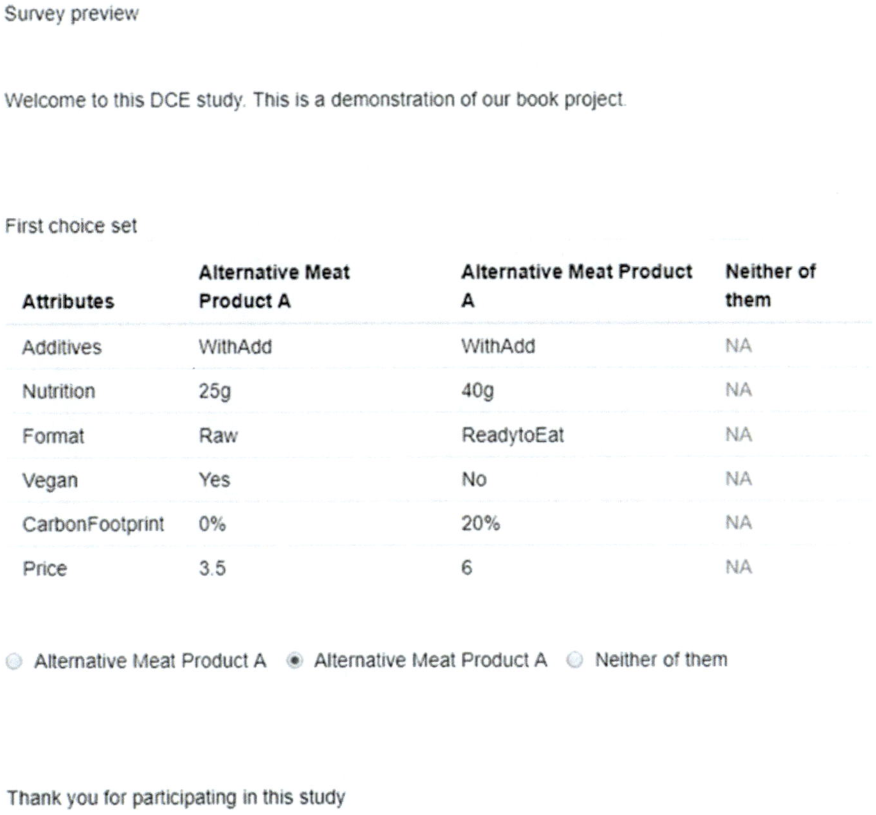

Fig. 8.5 Survey preview in DCEtool package

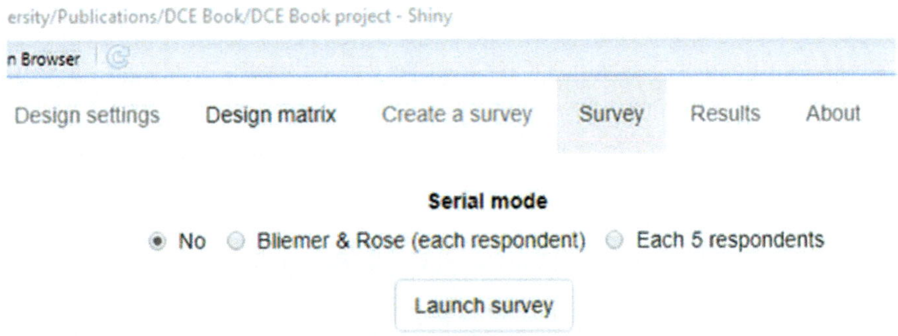

Fig. 8.6 Choosing a serial mode in DCEtool package

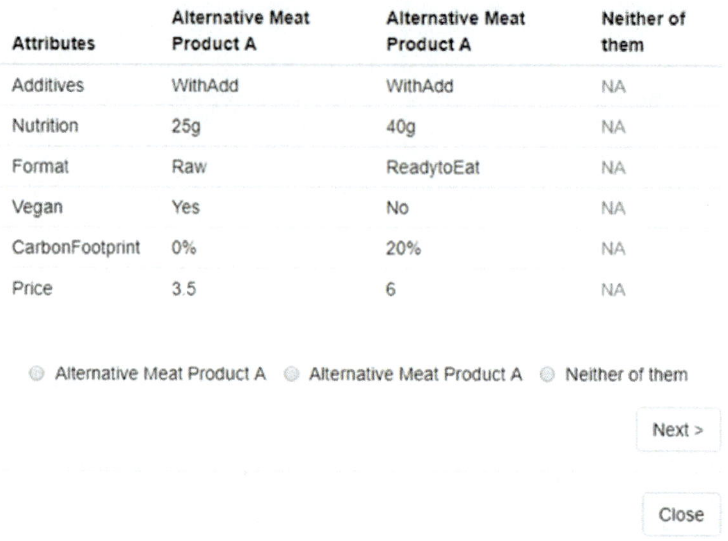

Attributes	Alternative Meat Product A	Alternative Meat Product A	Neither of them
Additives	WithAdd	WithAdd	NA
Nutrition	25g	40g	NA
Format	Raw	ReadytoEat	NA
Vegan	Yes	No	NA
CarbonFootprint	0%	20%	NA
Price	3.5	6	NA

Fig. 8.7 Launching a local survey in DCEtool

"**Launch survey**", a new pop-up window will appear, allowing respondents to evaluate DCE choice sets one by one and indicate their preferences (see Fig. 8.7 for an example).

8.2 From DCE Choice Sets to a Complete Questionnaire

Once the choice sets are finalized, our data collection needs to be organized through a survey questionnaire. Usually, DCE choice sets are only parts of a larger survey that contains at least five sections, as shown in Fig. 8.8. We will introduce each section in detail.

8.2.1 Introduction of the DCE Questionnaire

At the beginning of a DCE questionnaire, a non-technical introduction about the objective(s) and procedures of a DCE study should be clearly presented to respondents. Since DCEs involve scenarios assessments in which respondents' stated preferences rather than revealed preferences will be elicited, it is important that we specify the context of the study (e.g., what is the product/service/use context to be evaluated), what the respondents will be asked to do in the questionnaire, the reason for the selection of the respondents (e.g., the relevance of the study topic to the respondents), and the importance of their participation (e.g., how their response in

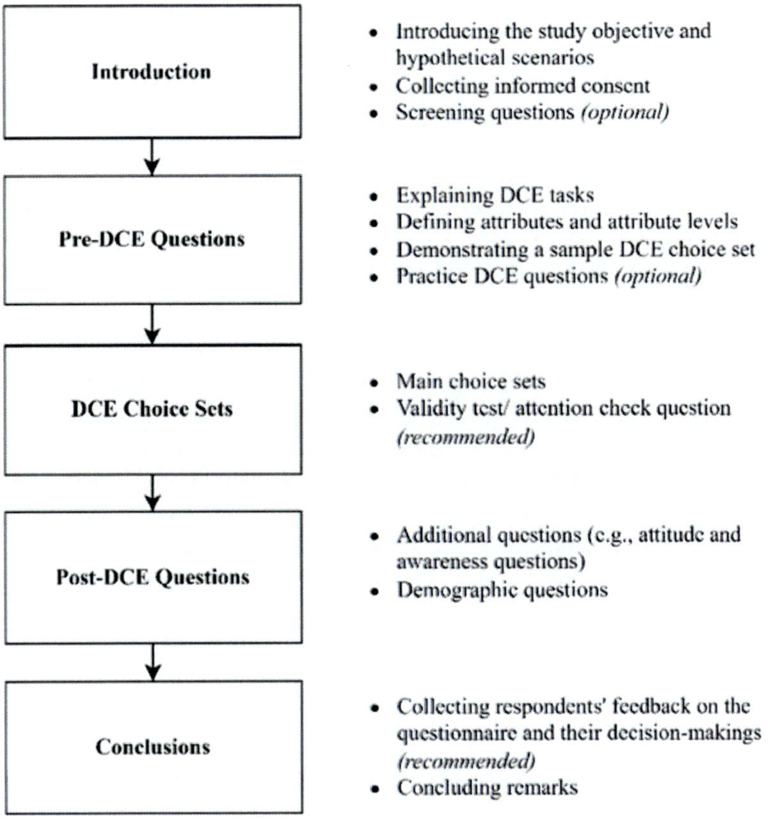

Fig. 8.8 DCE questionnaire structure

the questionnaire will enhance our understanding of public preference for medical treatments or public services or their wellbeing). Using our demonstrative project on alternative meat, for example, we show an introductory preface (see Textbox 8.1).

Textbox 8.1. A Sample Introductory Preface of a DCE Questionnaire
Alternative Meat Survey

Dear Participant,

We, researchers from XXX, are conducting a survey on people's perceptions of alternative meat. Alternative meat is a type of meat made of plants and other ingredients—but not from animals. The aim of this survey is to understand your opinion toward alternative meat. In this survey, you will be asked to indicate your preferences toward different alternative meat products. The survey is estimated to take approximately 10 to 15 minutes to complete. We have included an attention check question in the survey. Only those who pass the attention check will be directed to complete the survey.

Your participation in this survey is voluntary. You may refuse to take part in the research or exit the survey at any time without penalty.

You will receive no direct benefits from participating in this research study. However, your participation could help researchers learn more about public attitudes and preferences toward alternative meat products.

Your survey answers will be stored and kept safely by the research team at the university in a password-protected electronic format. This research has received ethics approval from the university: XXXXX.

If you have questions concerning the study, contact the principal investigator by email at XXXXX.

Informed consent should also be collected after presenting the introductory preface of a DCE study. Moreover, depending on research questions and research focus, some DCE studies may include *pre-screening* questions to exclude or disqualify respondents from answering the questionnaire. For instance, a DCE study examining patients' preferences for various treatments may *exclude* non-patients from participating in the study. Or a DCE study on entrepreneurs' preference for social value creation will exclude non-entrepreneurs from the screening process.

8.2.2 Pre-DCE Questions in the DCE Questionnaire

The framing of a DCE study design is very important as it determines the data quality to be collected. Before presenting a DCE study to respondents, a *detailed description* of the *attributes and attribute levels* that describe the topic under investigation, the *context* of the study, and the *overall choice scenarios* should be provided to respondents (Bennett & Blamey, 2001). We also suggest researchers include an example choice set or even several "practice" DCE questions for the respondents to familiarize themselves with before they start answering the actual choice tasks.

Using our demonstrative project on alternative meat, we present a sample description of a pre-DCE questionnaire (see Textbox 8.2).

Textbox 8.2. A Description of the Study Context
Before we start, please read the information below

Alternative Meat is made of non-animal-based proteins. Alternative meat products aim to mimic the appearance, texture, and taste of real (animal-based) meat. Alternative meat can be found in various products such as patties, cold cuts, and sausages.

In this questionnaire, we will use the **burger meat patty** as an example to study your attitude and preference toward alternative meat products. In this section, you will be presented with a choice task that consists of two imaginary alternative meat products for you to indicate your preference. Each product has a number of characteristics such as its nutrition, format, labeling price, etc. We want you to think about each product as if you were at a point of deciding whether to buy them in real life (e.g., in the supermarket). We will ask you to indicate which product you prefer to buy given the options that are available (e.g., product A or B). If both A and B are NOT attractive to you, you can choose "Neither of Them".

To help respondents warm up and better comprehend the rationale behind the choice tasks, we include a detailed explanation of the DCE attributes and their levels (using the alternative meat project as an example) after presenting an example choice set to respondents (see Textbox 8.3).

Textbox 8.3. A Sample Choice Set and Description of Attributes
A sample choice set is presented below.

Attributes	Alternative meat product A	Alternative meat product B
Additives	Without additives	With additives
Nutrition	25 grams proteins	40 grams proteins
Format	Raw	Ready-to-eat
Vegan label	With vegan label	Without vegan label
Carbon footprint	No info on reduction in greenhouse gas emission	20% reduction in greenhouse gas emission
Price	US$3.5	US$6

Which product do you prefer to buy?

- Product A
- Product B
- Neither of them

Choice Task Explanation:

As you can see above, each product is packaged as a 200-gram of alternative meat and is made up of 6 different characteristics: additives, nutrition, format, price, carbon footprint, and vegan label; and each characteristic has two or more specific types. Your job in the choice task is to evaluate each product based on its different characteristics and types and then decide which product you prefer to buy.

The product characteristics and their specific types are explained further below:
Additives here refer to whether additives are used to improve the product's flavor or appearance, or to keep it fresh or to prevent it from decay. Additives have two types which are Without additives and With additives.
Nutrition here refers to the "good and healthy thing" that you put in your body and we use "protein" as the definition for nutrition. Nutrition has two types which are 40 grams of proteins and 25 grams of proteins in the product.
Format here refers to whether the product requires any preparation before consumption, or whether it can be consumed right away. Format has two types which are Raw and Ready-to-eat.

Vegan Label refers to a label that indicates whether the product contains any real meat or not. Real meat includes anything from red meat, poultry, seafood, or any parts of an animal. Vegan label has two types which are Without vegan label and With vegan label.
Carbon Footprint here refers to the overall amount of greenhouse gases produced by a product, such as fossil fuel used to transport products, the pollution from growing the animal, etc. Carbon footprint has two types which are No info on reduction in greenhouse emission and 20% of reduction in greenhouse gas emission.
Price refers to the amount of money that one must pay to buy the product. Price has three types which are US$3.5, US$4.5, and US$6.

8.2.3 Presenting DCE Choice Tasks in a Questionnaire Format

The next step is to present the main DCE choice tasks to respondents and collect their preferences under different choice scenarios. If a DCE design only contains *one block*, meaning *all respondents will receive the same choice tasks*, we only need one section for the main DCE section. However, in some studies that involve a large number of choice tasks, we can consider blocking by partitioning the experimental design into multiple blocks that contain a limited number of choice tasks for each respondent (Weber, 2021) (also see Sect. 7.3 in Chap. 7 about the blocked designs).

For demonstration, we use the **blocked design** for our alternative meat demonstrative project and divide the entire 24 choice sets into a **fractional factorial study** design generated by the package **support.CEs** (see Fig. 7.3 in Chap. 7) into *three blocks, each with eight choice sets*. To implement this, the DCE should have *three parts*, and respondents will be *randomly directed to one of these three parts* (this can be easily done using Qualtrics). To randomize respondents across different blocks, we create and implement our DCE survey in Qualtrics, an online survey platform for creating and distributing web-based surveys. Qualtrics has a function that allows researchers to add a "*Randomizer*" in the survey flow to create branches that will display different blocks of choice tasks based on the conditions defined by researchers (Weber, 2021). A screenshot of the survey flow design in Qualtrics is shown in Fig. 8.9.

Another important consideration in the DCE design is whether to include a *validity check* to test for biases and the validity of a DCE experiment. For example, suppose we want to test respondents' rationality in their choice behavior (i.e., whether their choice behavior violates common rationality axioms). In that case, we could include a *dominance test* by including an additional choice set in the questionnaire. In the dominance test, *one of the choice alternatives should be designed in a way that is clearly superior to the other*, based on researchers' prior assumptions on attribute level ordering. Therefore, respondents who do not choose the superior alternative are

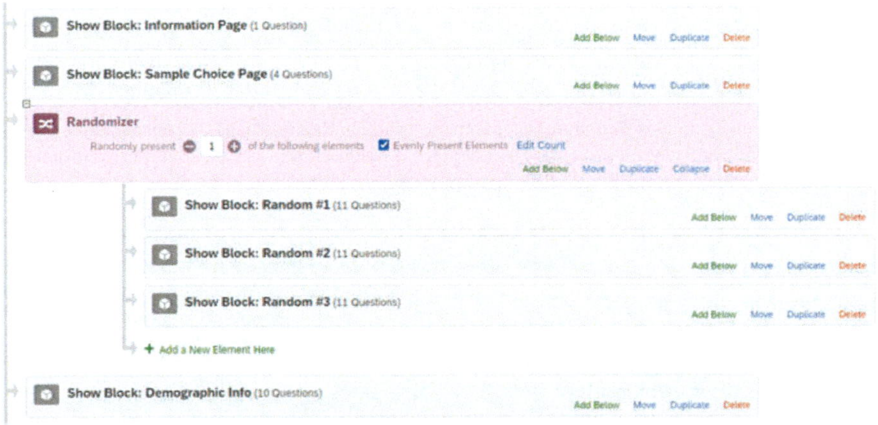

Fig. 8.9 Randomization design in three blocks using Qualtrics

considered to have failed the dominance test (Tervonen et al., 2018). This is similar to the pre-test used in a conventional experimental design where researchers manipulate certain variables and then screen out participants who clearly choose experimental conditions or choices that are obviously less favorable.

When respondents fail to choose the superior alternative in the choice tasks, it is very likely that they lack understanding of the choice tasks involved in the study, have not paid sufficient attention to the questionnaire, or simply randomly express their choices without much thinking. In most prior studies, respondents who fail the dominance test are excluded from data analysis (e.g., Chen et al., 2015; Finkelstein et al., 2015).

Using our demonstrative project on alternative meat to conduct a dominance test, we designed a choice set in which one alternative dominates the other on some attributes, such as price and nutrition, while the remaining attributes are the same. An example dominance test is shown in Table 8.2.

Table 8.2 Using a dominance test in a DCE study

Attributes	Alternative meat product A	Alternative meat product B
Additives Nutrition Format vegan label Carbon footprint Price	No additives **40 grams proteins** Ready-to-go Has vegan label 20% reduction in greenhouse gas emission **3.5 USD**	No additives **25 grams proteins** Ready-to-go Has vegan label 20% reduction in greenhouse gas emission **6 USD**
Which product do you prefer to buy?		
Product A ☐	Product B ☐	Neither of them ☐

Table 8.3 Using an attention check question in a DCE study

Attributes	Alternative meat product A	Alternative meat product B
Additives Nutrition format Vegan label Carbon footprint Price	Without additives 25 grams proteins Ready-to-eat Without vegan label No info on reduction in greenhouse gas emission US$4.5	With additives 40 grams proteins raw With vegan label 20% reduction in greenhouse gas emission US$6
Which product above has 40 grams proteins?		
Product A ☐	Product B ☐	

The only differences between alternative meat products A and B are their *nutrition and price*, in which Product A contains higher nutritional values but at a lower price compared to Product B. Therefore, respondents who have a clear understanding of the choice task should choose Product A over Product B.

However, in some situations, defining dominating alternatives might not be possible since attribute levels could not be ranked clearly based on preferences. Or in some other situations, respondents might infer additional information on attribute value beyond the information presented in the DCE choice sets. For example, respondents may infer higher prices as a signal of higher product quality for alternative meat products, resulting in choice behavior that contradicts researchers' prior expectations of what constitutes a rational choice (Tervonen et al., 2018).

Alternatively, we could also use *attention check* questions in the DCE study to screen out participants who have not paid enough attention to DCE choice tasks rather than using a dominance test. Attention check aims to identify unmotivated, reckless respondents who do not read the choice questions carefully could compromise data quality. For our demonstrative project, we include an attention check, as shown in Table 8.3.

8.2.4 Post-DCE Questions in the Questionnaire

After getting respondents' responses from the DCE questions (see Sect. 8.2.3 above), we asked post-DCE questions for additional data, for instance, the demographic characteristics of respondents, to further test respondents' understanding of the DCE choice tasks, and probe for deeper insights to understand respondents' rationale for making the choices in the way that they did (Bennett & Blamey, 2001). Below are

Fig. 8.10 Visualizing choice alternatives in a DCE study

some potential post-DCE questions that can be included in the questionnaire (of course, we should also be careful with the questionnaire length):

- **Further examination of preferences**: we could include a ranking or a rating exercise that asks respondents to evaluate each attribute (e.g., additives, nutrition) for a better understanding of the rationale behind respondents' choice behavior.
- **The degree of difficulty** in answering the DCE questionnaire: we could ask further questions to examine how difficult it was for respondents to answer the DCE questions and the parts that are considered difficult.
- **Socio-demographic questions**: we can and should include demographic questions such as age, gender, monthly income level, educational level, occupation, marital status, and so forth. These could serve as control variables in the study. They will also allow us to perform subgroup differences and preference heterogeneity.

A sample questionnaire is attached in Appendix I. To enhance respondents' understanding of the DCE study objectives, we created visuals (packaging of the alternative meat) that contain each alternative. As shown in Fig. 8.10, we modified the details on the product packaging (e.g., vegan label, nutritional value, carbon footprint label, additives, format) to present the variations in different attributes (Böttger et al., 2015). Researchers can use any software to manipulate the images to suit their study context.

8.3 Data Collection in DCE

8.3.1 Sampling and Sample Size

Before distributing DCE questionnaires, we should first determine our *sample frame*, which will be used to recruit respondents. The selection of the sampling frame depends on the nature of each DCE study as well as the research question(s). For instance, in our demonstrative project on "alternative meat", we aim to examine individuals' preferences for alternative meat products. An appropriate sampling frame to suit this question would be the general public who has heard of or consumed alternative meat products before. If our study aims to examine vegetarians' preference for alternative meat, then the most appropriate sampling frame would be vegetarian consumers. Examples of other sampling frames include people with disabilities, cancer patients, supermarket customers, tourists, family caregivers, disabled entrepreneurs, impact investors, etc.

After the sampling frame is known and decided upon, we could then determine the *sampling strategy* and *sample size*. For DCE studies, there is no specific requirement of what sampling strategies to adopt when implementing questionnaires. We can utilize *simple random sampling*, in which each respondent's probability of being selected is the same as every respondent in the sampling frame (Hoyos, 2010). Alternatively, we can use *stratified sampling* strategies by dividing the sampling frame into different groups based on certain criteria or variables (e.g., age, gender, occupation, performance). Each group represents a portion of the population (Louviere et al., 2000). For instance, a DCE study by Turner et al. (2007) divided the population into four groups by age (18–29, 30–54, 55–74, 75 + years) to examine their preferences toward the continuity of care in general practice.

While the above relies on probability sampling, DCE researchers also have used *non-probability sampling* strategies such as *purposive sampling* (e.g., Abiiro et al., 2014), *convenience sampling* (e.g., Oliver et al., 2019), *quota sampling* (e.g., Himmler et al., 2021), and *voluntary sampling* (e.g., Phillips et al., 2021) strategies.

Another key question is determining the *sample size* so the study has sufficient *statistical power* to detect a difference in respondents' preferences. In practice, the determination of sample size largely depends on the experimental design (e.g., the number of choice tasks that each respondent needs to answer, the number of alternatives in each choice task, and the number of blocks).

Bennett and Blamey (2001) suggested that the minimum size sample in each block should be at least 50. In other words, if our DCE design has three blocks, each block should have a minimum size of 50 respondents. De Bekker-Grob et al.'s (2015) review of sample size in DCE studies found three commonly used formulas to estimate the minimum sample size. These are the rules of thumb proposed by (1) **Johnson and Orme** (Orme, 1998); (2) **Pearmain et al.** (Pearmain et al., 1991); and (3) **Lancsar and Louviere** (Lancsar & Louviere, 2008).

In general, the sample size formula proposed by Johnson and Orme (Orme, 1998)—perhaps the most popular formula for sampling size determination—states that an effective sample size depends on the design factors, the total number of choice tasks, and the maximum number of attribute levels in the experimental design. The minimum sample size for DCE studies can be calculated using this formula (Orme, 1998):

$$N > \frac{500c}{(t * a)}$$

In this formula, N is the minimum sample size that we should have in a DCE study; c represents the largest number of attribute levels in our experimental design; t is the number of choice tasks, and a refers to the number of alternatives in each choice task. Using our demonstrative project on alternative meat as an example, a minimum of *94* respondents (we have 8 choice tasks, each with 2 alternatives, and the largest number of attribute levels is 3) is required. Orme (2010) suggested that the sample size formula was intended to calculate the minimum sample size and researchers should try to double the calculated minimum sample size at least.

In contrast, Pearmain et al. (1991) suggested that we should have a minimum of 100 respondents regardless of the experimental design in order to provide a basis for modeling preference data in DCE studies, while Lancsar and Louviere (2008) suggested a minimum of 10 observations (i.e., an observation refers to a response from a single participant to a single choice set[5]) per attribute plus 50 could have a sufficient sample size. Hensher et al. (2005) also suggested a formula of 50 respondents per attribute level. This means that if the DCE design includes five attributes, each with three levels, the minimum sample size would be $50 \times 3 \times 5 = 750$ respondents. The rationale behind this approach is to ensure enough variation in the attributes to capture the preferences of the population being studied.

Building upon these rules or formulas for sample size, De Bekker-Grob et al. (2015) introduced a new approach to determine the minimum sample size requirements for DCE studies. The calculation can be performed using the following five elements (de Bekker-Grob et al., 2015, p. 376):

- Significance level (α)
- Statistical power level ($1 - \beta$)
- The statistical model used in the DCE analysis [e.g., multinomial logit (MNL) model, mixed logit (MIXL) model, generalized multinomial logit (G-MNL) model][6]
- Initial belief about the parameter values (i.e., the relative importance of each attribute in influencing the choice)
- The DCE experimental design(e.g., the number of choice sets, the number of alternatives, the number of attributes and levels).

For calculating the minimum sample size, researchers will need first to determine the significance level, set the statistical power level, choose the statistical model to analyse the DCE data, establish initial beliefs about the parameter (attribute) values, and determine the DCE design.

Using our sample project (i.e., alternative meat products) for illustration, we can follow the steps outlined below to calculate the required sample size:

1. *Determine the number of choice sets*: suppose we adopt a fractional factorial design with 18 choice sets for each respondent;

[5] For example, if a respondent is presented with 1-choice sets provides a response to each one, then that would be considered 10 observations.

[6] For our sample project, we used the MNL model for estimation.

2. *Determine the initial belief about parameter values*: this depends on the prior studies. For the purpose of this example, we will assume that there is no prior knowledge about the parameter values;
3. *Determine the significance level* (α) *and statistical power level* ($1-\beta$): The significance level and statistical power level will depend on the specific research question and study design. For the purpose of this example, let's assume a significance level of 0.05 and a statistical power level of 0.80.

Using these factors, we can calculate the required sample size using the following equation:

$$n = [(z\alpha/2 + z\beta)^2 x P x (1 - P)]/(d^2 x r)$$

where:

- n is the required sample size.
- $z\alpha/2$ is the critical value for the selected significance level $\alpha/2$ (in a two-tailed test). For a significance level of 0.05, the critical value of $z\alpha/2$ is 1.96.
- $z\beta$ is the critical value for the selected statistical power level ($1-\beta$. For a statistical power level of 0.8, the critical value of $z\beta$ is 0.84.
- P is the proportion of respondents who choose the preferred option in the DCE. Assuming a two-tailed test, the proportion of respondents who choose the preferred option is 0.5.
- d is the minimum detectable difference in preference between two options, which is determined based on the standard deviation of the coefficients in the MNL model. We set d to 0.5 in this sample project.
- r is the number of choice sets per respondent in the DCE. We suppose there are 18 choice sets in this sample project.

Therefore, for our sample project:

$$n = [(1.96 + 0.84)^2 \times 0.5 \times (1 - 0.5)]/(0.1^2 \times 18)$$
$$n = 153.6$$

Therefore, we would need a minimum sample size of 154 respondents to ensure sufficient statistical power for our sample DCE study using an MNL model for estimation. More details on how to calculate the minimum sample size can be found in the paper by de Bekker-Grob et al. (2015).

8.3.2 DCE Questionnaire Administration

After the sampling strategy and the minimum sample size have been determined, the next step is implementing the DCE questionnaire so respondents can respond.

At this stage, researchers need to consider the data collection procedure. There are three main approaches for data collection in DCE studies:

- **Face-to-face interview**
- **Telephone interviews**
- **Web-based administration**

Face-to-face interviews allow researchers to explain and clarify the scenarios to respondents in more detail and to elaborate on them if they do not understand the choice task. This approach can be very helpful when participants come from *vulnerable groups* such as cancer patients, senior adults, disabled people (e.g., Green & Gerard, 2009; Kamphuis et al., 2015), or people with limited literacy (e.g., poor farmers from less developing worlds or uneducated refugees). However, this approach can be very costly and time-consuming, and it may bring interview bias where the interviewers may influence how respondents choose preferred alternatives (Ryan & Gerard, 2003). Alternatively, researchers can also use *telephone interviews* in which the interviewer and interviewees can have interactions without seeing each other in person. One drawback of this approach is that the questionnaire administration will highly rely on the oral presentation of the interviewer via telephone (Hoyos, 2010).

Web-based administration is relatively cheaper and most efficient compared to all available data collection approaches, and it allows researchers to present complex experimental designs in a simpler way. Moreover, the use of electronic questionnaires (or e-survey) allows randomizing of different blocks of DCE questions (for more complex DCE designs) or skipping questions (e.g., allowing patients who had never experienced a specific medical treatment to skip certain questions) and forcing respondents to answer all questions (Determann et al., 2017) (using "forced choice" mode).

In our demonstrative project of the alternative meat DCE study, we used a web-based administration where we implemented and distributed our DCE questionnaire via *Amazon Mechanical Turk* (MTurk; https://www.mturk.com/). MTurk is a paid survey platform that gives researchers access to consented participants and large samples of underrepresented groups that may be difficult to recruit through standard methods (Buhrmester et al., 2011). In our demonstrative project, we offered remuneration for MTurkers in exchange for their time in the study.

Appendix I: An Example of a DCE Questionnaire

Alternative Meat Survey
Dear Participant,

We, researchers from XXX, are conducting a survey on people's perceptions of alternative meat. Alternative meat is a type of meat made of plants and other ingredients—but not from animals. The aim of this survey is to understand your opinion towards alternative meat. In this survey, you will be asked to indicate your preferences towards different alternative meat products. The survey is estimated to take approximately 10 to 15 minutes to complete. We have included an attention check question in the survey. Only those who pass the attention check will be directed to complete the survey.

Your participation in this survey is voluntary. You may refuse to take part in the research or exit the survey at any time without penalty.

You will receive no direct benefits from participating in this research study. However, your participation could help researchers learn more about public attitudes and preferences toward alternative meat products.

Your survey answers will be stored and kept safely by the research team at the university in a password-protected electronic format. This research has received ethics approval from the university: XXXXX.

If you have questions concerning the study, contact the principal investigator by email at XXXXX.

Information Page

Alternative Meat or Meat Alternatives are on-animal-based protein that are usually constituted in a meat alternative product. Those products attempt to simulate the appearance, texture and taste of real meat. Alternative meat products can be in various formats such as patties, crumbles, or sausages. Some examples, which are entirely plant-based, include the Beyond meat patty and Omnipork luncheon below.

In this questionnaire, we will use the burger meat patty as an example to study consumers' preferences toward alternative meat products.

In this section, you will be presented with two imaginary "products". Each food product contains its own characteristics. We would like you to think about each product as if you were buying it and making a decision between them in the real world. Then we will ask you to tell us which product A or B you prefer more. You can also state that you do not prefer either product by choosing "Neither of Them". A sample choice set is presented below.

Attributes	Alternative Meat Product A (200 grams)	Alternative Meat Product B (200 grams)

Additives	WITHOUT additives	WITH additives
Nutrition	25 grams protein	40 grams protein
Format	Raw	Ready-to-Eat
Vegan Label	With vegan label	No vegan label
Carbon Footprint	No info on how the product reduces greenhouse gas emission	20% reduction in greenhouse gas emission
Price	\$3.5	\$6
Which product do you prefer to buy?		
Product A ☐ Product B ☐ Neither of Them ☐		

As you can see, each product is packaged as a 200-gram of alternative meat and is made up of 6 different attributes: additives, nutrition, format, price, carbon footprint, and vegan label; and each attribute has two or more specific types.

Your job in the choice task is to evaluate each product based on their different attributes and types and then decide which product you prefer to buy. The characteristics of products and their specific types are explained further as below:

The characteristics of products and their specific types are explained further as below:

- **Additives** refer to whether they are used to improve the product's flavor or appearance, or to keep it fresh or prevent it from decay. There are two types of additives: ***Without additives*** and ***With additives***.
- **Nutrition** refers to the "good and healthy thing" that you put in your body and we use "protein" as the definition for nutrition. There are two types of nutrition: ***40 grams*** of protein and ***25 grams*** of protein in the product.
- **Format** refers to whether the product requires any preparation before consumption, or whether it can be consumed right away. There are two types of formats: a "***Ready-to-Eat***" meal and ***Raw.***
- **Vegan Label** refers to a label that indicates whether the product contains any real (animal) meat or not. Real meat includes red meat, poultry, seafood, or any parts

of an animal. There are two types of vegan label: ***Without vegan label*** and ***With a vegan label***.

- **Carbon Footprint** refers to the overall amount of greenhouse gases produced by a food product, such as fossil fuel used to transport products, the pollution from growing the animal, etc. There are two types of carbon footprint: ***No info*** on how the product reduces greenhouse gas emission and The product uses ***20% less greenhouse emission*** in greenhouse gas respectively.

Price refers to the amount of money that one must pay to buy the product. There are three types of prices: ***$3.5***, ***$4.5***, and ***$6***.

In the following section, 8 choice sets will be presented to you. Please select which product (Product A or Product B) that you prefer the most.

1. Take a look at the two alternative meat products. Which one do you prefer?

Attributes	Alternative Meat Product A	Alternative Meat Product B

Additives	Without additives	With additives
Nutrition	25 grams proteins	40 grams proteins
Format	Raw	Ready-to-eat
Vegan Label	Without vegan label	With vegan label
Carbon Footprint	No info on reduction in greenhouse gas emission	20% reduction in greenhouse gas emission
Price	USD 3.5	USD 4.5

Which product do you prefer to buy?

Product A ? Product B ? Neither of them ?

2. Please have a look at this set of two alternative meat products. Which product do you prefer?

Attributes	Alternative Meat Product A	Alternative Meat Product B

Additives	Without additives	With additives
Nutrition	40 grams proteins	25 grams proteins
Format	Ready-to-eat	Raw
Vegan Label	With vegan label	Without vegan label
Carbon Footprint	No info on reduction in greenhouse gas emission	20% reduction in greenhouse gas emission
Price	USD 4.5	USD 6

Which product do you prefer to buy?

Product A ?   Product B ?   Neither of them ?

#ATT Check Question. Please have a look at this set of two alternative meat products. Please pick the one that contains 25 grams of proteins.

Attributes	Alternative Meat Product	Alternative Meat Product B

Additives	With additives	With additives
Nutrition	40 grams proteins	25 grams proteins
Format	Raw	Raw
Vegan Label	Without vegan label	Without vegan label
Carbon Footprint	No info on reduction in greenhouse gas emission	No info on reduction in greenhouse gas emission
Price	USD 6	USD 3.5

Which product contains 25 grams proteins?

Product A ?   Product B ?

##If respondents select A, this means that they fail to understand the task. We will then redirect them to the end of the survey. If they select B, they will be allowed to continue the survey##

Ending message for those who do not pass the attention test:

We are sorry that you do not pass the attention check. The survey will be closed now.

3. Please have a look at this set of two alternative meat products. Which product do you prefer?

Attributes	**Alternative Meat Product A**	**Alternative Meat Product B**

Additives	With additives	Without additives
Nutrition	25 grams proteins	40 grams proteins
Format	Ready-to-eat	Raw
Vegan Label	Without vegan label	With vegan label
Carbon Footprint	No info on reduction in greenhouse gas emission	20% reduction in greenhouse gas emission
Price	USD 3.5	USD 4.5

Which product do you prefer to buy?

Product A ? Product B ? Neither of them ?

4. Please have a look at this set of two alternative meat products. Which product do you prefer?

Attributes	Alternative Meat Product A	Alternative Meat Product B

Additives	Without additives	With additives
Nutrition	25 grams proteins	40 grams proteins
Format	Raw	Ready-to-eat
Vegan Label	With vegan label	Without vegan label
Carbon Footprint	20% reduction in greenhouse gas emission	No info on reduction in greenhouse gas emission
Price	USD 4.5	USD 3.5

Which product do you prefer to buy?

Product A ? Product B ? Neither of them ?

5. Please have a look at this set of two alternative meat products. Which product do you prefer?

Attributes	Alternative Meat Product A	Alternative Meat Product B

Additives	Without additives	With additives
Nutrition	40 grams proteins	25 grams proteins
Format	Raw	Ready-to-eat
Vegan Label	With vegan label	Without vegan label
Carbon Footprint	20% reduction in greenhouse gas emission	No info on reduction in greenhouse gas emission
Price	USD 3.5	USD 4.5

Which product do you prefer to buy?

Product A ? Product B ? Neither of them ?

6. Please have a look at this set of two alternative meat products. Which product do you prefer?

Attributes	Alternative Meat Product A	Alternative Meat Product B

Additives	Without additives	With additives
Nutrition	25 grams proteins	40 grams proteins
Format	Raw	Ready-to-eat
Vegan Label	Without vegan label	With vegan label
Carbon Footprint	20% reduction in greenhouse gas emission	No info on reduction in greenhouse gas emission
Price	USD 6	USD 3.5

Which product do you prefer to buy?

Product A ? Product B ? Neither of them ?

7. Please have a look at this set of two alternative meat products. Which product do you prefer?

Attributes	Alternative Meat Product A	Alternative Meat Product B

Additives	Without additives	With additives
Nutrition	25 grams proteins	40 grams proteins
Format	Ready-to-eat	Raw
Vegan Label	With vegan label	Without vegan label
Carbon Footprint	No info on reduction in greenhouse gas emission	20% reduction in greenhouse gas emission
Price	USD 4.5	USD 6

Which product do you prefer to buy?

Product A ? Product B ? Neither of them ?

8. Please have a look at this set of two alternative meat products. Which product do you prefer?

Attributes	Alternative Meat Product A	Alternative Meat Product B

Additives	Without additives	With additives
Nutrition	25 grams proteins	40 grams proteins
Format	Ready-to-eat	Raw
Vegan Label	Without vegan label	With vegan label
Carbon Footprint	No info on reduction in greenhouse gas emission	20% reduction in greenhouse gas emission
Price	USD 4.5	USD 6

Which product do you prefer to buy?

Product A ? Product B ? Neither of them ?

Demographic questions:

1. What is your gender?
   A. Male
   B. Female
   C. Others
2. What is your age?
   A. 18–25 years old
   B. 26–35 years old
   C. 36–45 years old
   D. 46–55 years old
   E. 56–65 years old
   F. 66–75 years old
   G. > 75 years old
3. What is your present religion, if any?
   A. Christian
   B. Muslim
   C. Jewish
   D. Buddhist
   E. Others (SPECIFY_______)
   F. Nothing in particular
4. What is your marital status?
   A. Single
   B. Married
   C. Divorced
   D. Widowed
5. In which country do you currently reside?
   Answer:
6. What is the highest level of education you have completed?
   A. Primary School or Below

B. Secondary School
C. Diploma
D. Associate Degree/High Diploma K. Bachelor's Degree

7. What is your current employment status?
   A. Student
   B. Employed Full-Time
   C. Employed Part-Time
   D. Temporary Worker
   E. Seeking opportunities
   F. Retired
   G. Others
8. What is your occupation (if applicable)?
9. What is your monthly income (in US dollars)?
   A. <500
   B. 500–1000
   C. 1001–2000
   D. 2001–3000
   E. 3001–4000
   F. >4000
10. What are your eating habits?
    A. Meat-eating
    B. Eat white meat only
    C. Pescatarian
    D. Vegetarian
    E. Vegan
    F. Other (please specify)________________

Additional Questions: Environmental concerns, behavioral intention, and attention to environmental impact of food choices

Please rate the extent to which you think each of the following statement describes you:

	Strongly agree (1)	Somewhat agree (2)	Neither agree nor disagree (3)	Somewhat disagree (4)	Strongly disagree (5)
I am very concerned about the environment. (1)	❍	❍	❍	❍	❍
I would be willing to reduce my consumption to help protect the environment (2)	❍	❍	❍	❍	❍
Major political change is necessary to protect the natural environment. (3)	❍	❍	❍	❍	❍
Major social changes are necessary to protect the natural environment. (4)	❍	❍	❍	❍	❍
Anti-pollution laws should be enforced more strongly. (5)	❍	❍	❍	❍	❍
I try to separate trash for recycling on a regular basis (6)	❍	❍	❍	❍	❍
I prefer to use environmental-friendly products (7)	❍	❍	❍	❍	❍
I want to use products that I currently have as long as possible (8)	❍	❍	❍	❍	❍
When I buy foods, I try to consider how my use of them will affect the environment (9)	❍	❍	❍	❍	❍

Questionnaire Ending

Thank you for participating in our research study, the team are very appreciated with your effort on the questionnaire in finding out customers preferences on Alternative Meat, your answers will help us a lot.

References

Abiiro, G. A., Leppert, G., Mbera, G. B., Robyn, P. J., & De Allegri, M. (2014). Developing attributes and attribute-levels for a discrete choice experiment on micro health insurance in rural Malawi. *BMC Health Services Research, 14*(1), 1–15.

Aizaki, H. (2012). Basic functions for supporting an implementation of choice experiments in R. *Journal of Statistical Software, 50*, 1–24.

Bennett, J., & Blamey, R. (Eds.). (2001). *The choice modelling approach to environmental valuation*. Edward Elgar Publishing.

Bliemer, M. C., & Rose, J. M. (2010). Serial choice conjoint analysis for estimating discrete choice models. In *Choice modelling: The state-of-the-art and the state-of-practice* (pp. 137–161). Emerald Group Publishing Limited.

Böttger, B., Thate-Waschke, I. M., Bauersachs, R., Kohlmann, T., & Wilke, T. (2015). Preferences for anticoagulation therapy in atrial fibrillation: The patients' view. *Journal of Thrombosis and Thrombolysis, 40*(4), 406–415.

Buhrmester, M., Kwang, T., & Gosling, S. D. (2011). Amazon's mechanical Turk: A new source of inexpensive, yet high-quality, data? *Perspectives on Psychological Science, 6*(1), 3–5.

Chen, L. C., Cheng, L. J., Zhang, Y., He, X., & Knaggs, R. D. (2015). Acupuncture or low frequency infrared treatment for low back pain in Chinese patients: A discrete choice experiment. *PLoS One, 10*(5), e0126912.

De Bekker-Grob, E. W., Donkers, B., Jonker, M. F., & Stolk, E. A. (2015). Sample size requirements for discrete-choice experiments in healthcare: A practical guide. *The Patient-Patient-Centered Outcomes Research, 8*(5), 373–384.

Determann, D., Lambooij, M. S., Steyerberg, E. W., de Bekker-Grob, E. W., & De Wit, G. A. (2017). Impact of survey administration mode on the results of a health-related discrete choice experiment: Online and paper comparison. *Value in Health, 20*(7), 953–960.

Finkelstein, E. A., Bilger, M., Flynn, T. N., & Malhotra, C. (2015). Preferences for end-of-life care among community-dwelling older adults and patients with advanced cancer: A discrete choice experiment. *Health Policy, 119*(11), 1482–1489.

Green, C., & Gerard, K. (2009). Exploring the social value of health-care interventions: A stated preference discrete choice experiment. *Health Economics, 18*(8), 951–976.

Hensher, D. A., Rose, J. M., Rose, J. M., & Greene, W. H. (2005). *Applied choice analysis: A primer*. Cambridge University Press.

Himmler, S., Soekhai, V., van Exel, J., & Brouwer, W. (2021). What works better for preference elicitation among older people? Cognitive burden of discrete choice experiment and case 2 best-worst scaling in an online setting. *Journal of Choice Modelling, 38*, 100265.

Hoyos, D. (2010). The state of the art of environmental valuation with discrete choice experiments. *Ecological Economics, 69*(8), 1595–1603.

Kamphuis, C. B., de Bekker-Grob, E. W., & van Lenthe, F. J. (2015). Factors affecting food choices of older adults from high and low socioeconomic groups: A discrete choice experiment. *The American Journal of Clinical Nutrition, 101*(4), 768–774.

Lancsar, E., & Louviere, J. (2008). Conducting discrete choice experiments to inform healthcare decision making. *Pharmacoeconomics, 26*(8), 661–677.

Louviere, J. J., Hensher, D. A., & Swait, J. D. (2000). *Stated choice methods: Analysis and applications*. Cambridge University Press.

Oliver, D., Deal, K., Howard, M., Qian, H., Agarwal, G., & Guenter, D. (2019). Patient trade-offs between continuity and access in primary care interprofessional teaching clinics in Canada: A cross-sectional survey using discrete choice experiment. *British Medical Journal Open, 9*(3), e023578.

Orme, B. (1998). Sample size issues for conjoint analysis studies. *Sequim: Sawtooth Software Technical Paper*.

Orme, B. K. (2010). Sample size issues for conjoint analysis: Getting started with conjoint analysis: Strategies for product design and pricing research. Madison, Wis. *Madison (WI): Research Publishers LLC*, 57–66.

Pearmain, D., Swanson, J., Kroes, E., & Bradley, M. (1991). *Stated preference techniques: A guide to practice* (2nd ed.). Steer Davies Gleave and Hague Consulting Group.

Perez-Troncoso. (2022). *Design, conduct and analyze 'DCEs' from a virtual interface in shiny*. Retrieved from https://github.com/danielpereztr/DCEtool

Phillips, E. A., Himmler, S. F., & Schreyögg, J. (2021). Preferences for e-mental health interventions in Germany: A discrete choice experiment. *Value in Health, 24*(3), 421–430.

Ryan, M., & Gerard, K. (2003). Using discrete choice experiments to value health care programmes: Current practice and future research reflections. *Applied Health Economics and Health Policy, 2*(1), 55–64.

Tervonen, T., Schmidt-Ott, T., Marsh, K., Bridges, J. F., Quaife, M., & Janssen, E. (2018). Assessing rationality in discrete choice experiments in health: An investigation into the use of dominance tests. *Value in Health, 21*(10), 1192–1197.

Traets, F., Sanchez, D. G., & Vandebroek, M. (2020). Generating optimal designs for discrete choice experiments in R: The idefix package. *Journal of Statistical Software, 96*, 1–41.

Turner, D., Tarrant, C., Windridge, K., Bryan, S., Boulton, M., Freeman, G., & Baker, R. (2007). Do patients value continuity of care in general practice? An investigation using stated preference discrete choice experiments. *Journal of Health Services Research & Policy, 12*(3), 132–137.

Weber, S. (2021). A step-by-step procedure to implement discrete choice experiments in Qualtrics. *Social Science Computer Review, 39*(5), 903–921.

Chapter 9
DCE Data Analysis Using R

In this chapter, we discuss and demonstrate the techniques to analyze the generated choice data, starting from data preparation, model selection, model specification, and estimation. This demonstration involves the use of various R packages for estimating the classical multinominal logit model, calculating the goodness of fit, and computing the marginal willingness to pay values (i.e., the relative importance of non-monetary attributes in DCE studies). In particular, we demonstrate the use of ***clogit()*** function in the R package **survival** for model estimation, as well as using the **MNL model, and** calculating the goodness of fit called **McFadden R^2,** estimating the willingness to pay (**WTP**) of non-monetary attributes, and i**nteraction effects analysis** among two or more attributes.

9.1 Model Selection

After collecting DCE data via a DCE survey using online survey, interviews, or other approaches, the next step is to implement choice modeling, the most critical element in DCE studies. DCE data often involves using regression models with a *dichotomous* or *polychotomous* dependent variable (a type of 'categorical variable') (Mangham et al., 2009). In its simplest form, we can define the observed utility of the product or service under study as a linear expression in which each attribute is weighted by a specific parameter[1] representing the marginal utility of that particular attribute. The probability of choosing one alternative over another in DCE studies can be described using the following formula (Hauber et al., 2016, p. 302), in which X_i is the attribute level of attribute *i*, β_0 is the intercept (also known as the alternative-specific constant; ASC), and β_i is the preference weight for attribute i:

[1] A parameter represents the marginal utility or importance of a specific attribute level on the overall choice.

L. Shang and Y. Chandra, *Discrete Choice Experiments Using R*,
https://doi.org/10.1007/978-981-99-4562-7_9

$$\Pr(\text{choice}) = \beta_0 + \sum\nolimits_i \beta_i X_i$$

This simple formula has the advantage of being very easy to estimate. However, it also has several limitations, such as the research must assume that each attribute has an *independent* and *linear effect* on preference (Roux et al., 2004). This formula assumes that the attribute levels are orthogonal to each other, meaning that they do not interact with each other, which may not be realistic for all attributes in all contexts. Another limitation of this formula is that it assumes that the parameters are constant across all individuals, which may not be the case in reality.

Moreover, there is a wide range of choice models to estimate the DCE data, which depends on various considerations. Various assumptions concerning the variance–covariance matrices of preference parameters, distributions, and properties of error components require using different choice models (Lancsar & Louviere, 2008).

The model selection can be based on economic and behavioral theories, the experimental design of DCEs, and statistical considerations (Kjaer, 2005). For example, the theory of rational choice suggests that individuals make choices based on their preferences and available information. In the context of DCE analysis, adopting this theory may lead to the use of a multinomial logit (MNL) model, as it assumes that individuals make choices based on the attributes of the alternatives and the relative importance of those attributes to the individual. The experimental design of DCEs can also influence model selection. For instance, if a DCE includes attributes with more than two levels or if there are interactions between attributes, a more complex model, like the mixed logit model, may be more appropriate.

Broadly speaking, there are two significant types of choice models for DCE analysis: **binary discrete choice models** (used when there are only one alternative and an opt-out option: a 'yes or no' vote) and **multiple discrete choice models** (used when there are three or more alternatives in each choice set) (Louviere et al., 2000; Train, 2009). We will briefly introduce these different choice models below.

The **binary discrete choice models** are the simplest discrete models in DCE studies. If the choice question asks whether a respondent accepts an option (or takes an action) or not, *binary logit* and *binary probit* models can be applied in the estimation, which are the two best-known binary discrete choice models. According to Kjaer (2005), logit models have closed-form expressions. They are built based on the assumption that error terms are independent of each other and are identically distributed. In contrast, the probit models assume a normal distribution in which the unobservable part of the utility is usually spread. In general, the choice between binary logit and binary probit models should be made based on the theoretical assumptions and interpretability of coefficients while taking into account the goodness of fit and computational considerations.

When there are *two or more alternatives* in each DCE choice set, we need to select a choice model from **multinomial discrete choice models**. Among various models, one commonly used model is the **multinomial logit (MNL)** (also known as the **conditional logit model**), developed by McFadden (1974). MNL is a *fixed effects logit* that relates the probability of choice among two or more alternatives to the features of the attribute levels that are used to describe the alternatives. This model assumes that individuals make rational choices based on the levels of each attribute and that the utility (or satisfaction) they derive from a particular alternative can be expressed as a linear function of those attributes. It is relatively easy to estimate compared to other models (Hauber et al., 2016). Moreover, McFadden (1974) has shown that the MNL model is consistent with the random utility theory. MNL model is rooted in the *independence of irrelevant alternatives* (IIA) property, which means that the ratio of the probabilities of preferring one alternative over the other will not be affected by adding any new alternatives in the choice sets. Mathematically, this can be represented as (Eq. 9.1):

$$\mathrm{P}(i)/\mathrm{P}(j) = \exp(\mathrm{V}_i - \mathrm{V}_j) \tag{9.1}$$

where P(i) and P(j) are the probabilities of choosing alternatives i and j, and V_i and V_j are the deterministic utilities of alternatives i and j, respectively.

The IIA property has both advantages and limitations. On the one hand, it simplifies the estimation of model parameters and allows the MNL model to be easily applied to various choice situations. On the other hand, however, the MNL model assumes that utility varies among alternatives but not among respondents, and thus, it assumes preference homogeneity and estimates the same coefficients for all respondents (Steckel & Vanhonacker, 1988). Moreover, whether IIA holds is also questionable (Mazzanti, 2003). For example, if a decision-maker's preferences are affected by the similarity between alternatives, the IIA assumption may not hold. Some models that relax IIA include the **nested logit model**, **multinomial probit**, and **mixed logit model**.

The **nested logit (NL)** model is an extension of the traditional MNL model. The opt-out alternative is often experienced and preferred by respondents, and they may consider the non-opt-out alternatives (e.g., Alternative Meat A vs. B in our demonstrative project) as substitutes for each other when participating in a DCE study. Consequently, the non-opt-out alternatives might share similar unobserved utilities, and thus their utilities are more correlated among themselves rather than with the opt-out alternatives (Bliemer et al., 2009; Campbell & Erdem, 2019). Therefore, the NL model suggests that respondents initially choose between "opt-out" (status quo situation or doing nothing) and doing something. If they decide "doing something" over remaining the same, they will subsequently select different alternatives, such as choosing between Alternative Meat A vs. B (Kjær, 2005).

In contrast, **multinomial probit** is a more complex model that relaxes IIA assumption and allows the analysis of correlation over alternatives and time using fixed error variance (de Bekker-Grob et al., 2012). In the MNL model, the correlation is restricted to zero, which assumes that the unobserved factors affecting the utility of one alternative are not related to the unobserved factors affecting the utilities of the other alternatives. The multinomial probit model, on the other hand, allows the error terms to be correlated across alternatives. Thus, this model can capture the fact that unobserved factors affecting the utility of one alternative may also affect the utilities of the other alternatives. However, it is computationally more complex and demanding (Hoyos, 2010).

Another popular model is the **mixed logit model (MXL)** (also known as the **random-parameters logit**), which relaxes the assumption of preference homogeneity and is not subject to IIA assumption (de Bekker-Grob et al., 2012). This model has grown in popularity in DCE studies over the past years. Hoyos (2010) highlighted that the MXL model could address most of the abovementioned disadvantages and limitations of other models and allow the estimation of choice probabilities of any discrete choice model based on the random utility theory (McFadden & Train, 2000). This model assumes the probability of choosing one alternative over others is based on the utilities of attribute levels that describe the alternatives plus a random error term that adjusts for individual-specific variances in preferences (Hauber et al., 2016; Kjær, 2005).

A wide range of R packages is available to support the estimation of different choice models. For instance, the R package **survival** has a function ***clogit()*** that allows the estimation of the MNL model. The package **mlogit** enables the estimation of the traditional MNL and also extended MNL model (e.g., the nested logit and the heteroscedastic model). In addition, the R package **ChoiceModelR** supports the estimation of a hierarchical multinomial logit model (also known as the hierarchical Bayes multinomial logit model) that account for unobserved heterogeneity and potential IIA violations. The **bayesm** package allows the estimation of the binary probit model through its function ***rbprobitGibbs()***.

This chapter will focus on the most popular model, MNL, and introduce various ways of using R to perform preference estimation analyses.[2]

9.2 Data Preparation for Analysis

DCE studies contain a data matrix where each row corresponds to a single profile in the DCE experiment. There are various ways to define the attribute levels in each row of data. For attributes with numerical levels (e.g., price, cost, level of risks), we can specify them as *continuous variables* and their level values will appear in the data set. For other attributes, we can specify them as categorical attributes.

There are three coding approaches for categorical attributes for DCE analysis, which are *mean-centering numerical coding*, *effects coding*, and *dummy-variable coding* (Hauber et al., 2016; Hoyos, 2010). The latter two coding approaches are more commonly used for categorical coding of attribute levels (Bech & Gyrd-Hansen, 2005; Hensher et al., 2005). For effects coding and dummy-variable coding approaches, one level of each attribute needs to be omitted (e.g., for a 3-level attribute, the first level will be omitted, and only the latter two levels will be presented), and

[2] To better illustrate the use of various R packages for analyzing DCE data, we created mock-up datasets rather than using real data for demonstration purposes.

each non-omitted attribute level is assigned a value of 1 when it is present. Otherwise, a value of 0 will be given.

The difference between effects coding and dummy-variable coding is that when the omitted level is present, all other non-omitted levels will be coded as -1 in effects coding. In contrast, they will be coded as 0 in the dummy-variable coding. Table 9.1 below shows the differences between the two approaches using our demonstrative project that examines consumers' preferences for alternative meat products:

For our demonstrative project on alternative meat, we administered the questionnaire via Qualtrics (an online survey platform) and collected data via MTurk (a paid survey platform). Then, we exported the choice data from Qualtrics into an Excel spreadsheet format. Before starting the preference estimation, we had to organize the data according to the requirements of R packages. Two elements are necessary to prepare our DCE data set: (1) our *experimental design* and (2) *responses* to DCE questions. Then, we can either organize the data set manually or use R functions to assist our data transformation.

For instance, the function ***clogit()*** in the R package **survival** (Therneau & Lumley, 2015) supports the estimation of the MNL model, but it requires the data set to be organized in a specific format in which each alternative comprises one row in the data set.

To prepare the DCE data set, we recommend researchers use a powerful R package called **support.CEs,** which has two functions to create a data set suitable for the model estimation function ***clogit().*** The first function is ***make.design.matrix(),*** which converts a DCE design we created using the function ***rotation.design()*** or ***Lma.design()*** (see Sect. 7.5.1 in Chapter 7 for more details) into a design matrix.

The second function is ***make.dataset()*** which combines a data set containing information about respondents' answers to the DCE questions and a data set containing a design matrix related to these DCE questions. These two functions allow us to create a data set that can be analyzed using the function ***clogit()*** (Aizaki, 2012).

Below are the steps to prepare for DCE dataset using support.CEs.

1. Extracting DCE Responses to Excel Format

First, we need to extract respondents' responses to DCE questions from the completed questionnaires and organize the responses in the following Excel format for further processing (see Fig. 9.1).

In the above spreadsheet, the column "ID" refers to *each respondent*, and the column "BLOCK" refers to which DCE block the respondent was assigned to (if the block design was not adopted in the experiment, there would only be one block or "block 1"). The columns "q1" to "q8" capture the respondents' answers to each DCE question (where they evaluate each choice set that contains two alternatives and an opt-out option), and a total of 8 choice sets (each with two alternatives) were included in our demonstrative DCE project.

The value of "1" means the respondent selected *Alternative Meat Product A* as their preferred option, the value of "2" means *Alternative Meat Product B* was chosen, and the value of "3" refers to the opt-out option.

Table 9.1 Effects coding and dummy-variable coding

	ID	X	Res	ASC	Additives_with	Nutrition_40g	Format_ready	Carbon_20%	Vegan_Yes	Price
Dummy-variable coding										
1	1	set1.alt1	1	0	1	1	1	0	0	45
2	1	set1.alt2	0	0	0	0	0	1	1	25
3	1	set1.alt3	0	1	0	0	0	0	0	0
4	1	set2.alt1	1	0	0	1	0	0	0	25
5	1	set2.alt2	0	0	1	0	1	1	1	35
6	1	set2.alt3	0	1	0	0	0	0	0	0
Effects coding										
1	1	set1.alt1	1	0	1	1	1	− 1	− 1	45
2	1	set1.alt2	0	0	− 1	− 1	− 1	1	1	25
3	1	set1.alt3	0	1	− 1	− 1	− 1	− 1	− 1	0
4	1	set2.alt1	1	0	− 1	1	− 1	− 1	− 1	25
5	1	set2.alt2	0	0	1	− 1	1	1	1	35
6	1	set2.alt3	0	1	− 1	− 1	− 1	− 1	− 1	0

	A	B	C	D	E	F	G	H	I	J
1	ID	BLOCK	q1	q2	q3	q4	q5	q6	q7	q8
2	1	1	1	1	1	1	2	1	2	2
3	2	1	2	1	2	1	1	2	1	1
4	3	1	2	1	1	1	2	2	3	3
5	4	1	1	1	1	1	2	1	2	2
6	5	1	2	1	1	1	2	2	2	1
7	6	1	1	1	1	1	2	1	2	2
8	7	1	1	1	1	1	2	1	2	2
9	8	1	1	2	3	1	1	3	1	2
10	9	1	2	2	1	1	2	2	1	1
11	10	1	2	2	2	1	2	2	2	1
12	11	1	2	1	1	1	2	2	2	2
13	12	1	2	1	1	1	2	2	2	1
14	13	1	1	1	1	1	2	1	1	2
15	14	1	2	1	2	1	2	2	1	1
16	15	1	1	2	1	1	2	1	1	1
17	16	1	1	1	1	1	2	1	2	2
18	17	1	2	1	1	1	2	1	2	3
19	18	1	2	1	2	1	2	2	2	2
20	19	1	2	1	2	1	2	2	1	1
21	20	1	2	1	2	1	1	2	1	2
22	21	1	2	1	2	1	2	2	1	2

Fig. 9.1 Organizing DCE questionnaire responses in excel

Please note that the Excel spreadsheet should be saved as a **CSV file** to smoothly import the data into R later.

2. Importing the DCE Data File to R

After preparing the CSV file, the next step is to import the file into R by using the following code:

```
df <- read.csv("d:\\test\\dataset_sample_DCE.csv", header = TRUE, as.is = TRUE)
```

Note that your data file can be stored in a computer with a different location than the "d" drive (as in the code above) and written in a file label that is different from "dataset_sample_DCE.csv" thus when importing the file to R, your R code must use the right file location and file name.

3. Data Transformation for DCE Analysis Using Clogit()

Next, we can convert the data file into a dataset suitable for the function ***clogit()*** by using the functions ***make.design.matrix()*** and ***make.dataset()*** in the R package **support.CEs**:

```
library(support.CEs)

desmat2 <- make.design.matrix(choice.experiment.design = unlabeled, optout = TRUE,
                    categorical.attributes = c("Additives", "Nutrition", "Convenience",
"CarbonFootprint", "VeganLabel"),
                    continuous.attributes = c("Price"),
                    unlabeled = TRUE)

dataset2 <- make.dataset(respondent.dataset = df,
                 choice.indicators = c("q1", "q2", "q3", "q4", "q5", "q6", "q7", "q8"),
                 design.matrix = desmat2)
```

unlabeled is the object where we saved our experimental design previously. Please refer to section 7.5.1. in Chapter 7 for more information

By executing the R code above, we can link participants' responses to each choice set with our experimental design. The converted data set is printed on the R console as below (see Fig. 9.2). For instance, lines 1–24 show the responses of the first participant (with ID #1) to the eight choice sets (each with two alternatives and one opt-out option) we have in the experiment. Here is a brief explanation of each column:

- ID: The respondent ID
- BLOCK: The block number (if the experiment was designed in blocks)
- QES: The question number (if the experiment has multiple questions)
- ALT: The alternative number within the choice set
- RES: Whether the alternative is the respondent's chosen option (1) or not (0)
- ASC: Whether the option is an alternative (1) or an opt-out option (0).
- Yes, Forty, Ready, Twenty, Yes.2: The levels of the categorical attributes (as specified in the *categorical.attributes* argument of the ***make.design.matrix()*** function)
- Price: The value of the continuous attribute (as specified in the *continuous.attributes* argument of the ***make.design.matrix()*** function)
- STR: A string of characters that represents the choice indicators (as specified in the *choice.indicators* argument of the ***make.dataset()*** function). In this case, the indicators are "q1", "q2", "q3", "q4", "q5", "q6", "q7", "q8".

Similarly, the R package **idefix** also supports data transformation through its function *Datatrans* and allows researchers to convert the data set into formats that other packages can analyze. To use this function, we will need first to extract questionnaire data and convert the **"wide" data format** (as shown in Fig. 9.1 where the 8 choice sets are the columns and each choice set has a value of either 1 or 2 or 3) to a **"long" data format** in which each row refers to a profile or "alternative" in a choice set and shows the value of respondents' response to that profile (whether that profile is selected as the preferred option or not, which is represented as either "0" or "1").

We can use the R package **tidyverse** to help reshape our exported questionnaire data:

```
R 4.2.0 · D:/OneDrive - Lingnan University/Publications/DCE Book/DCE Book project/
> dataset2
   ID BLOCK QES ALT   RES ASC Yes Forty Ready Twenty Yes.2 Price STR
1   1     1   1   1  TRUE   1   1     1     1      0     0    45 101
2   1     1   1   2 FALSE   1   0     0     0      1     1    25 101
3   1     1   1   3 FALSE   0   0     0     0      0     0     0 101
4   1     1   2   1  TRUE   1   0     1     0      0     0    25 102
5   1     1   2   2 FALSE   1   1     0     1      1     1    35 102
6   1     1   2   3 FALSE   0   0     0     0      0     0     0 102
7   1     1   3   1  TRUE   1   0     0     1      0     0    35 103
8   1     1   3   2 FALSE   1   1     1     0      1     1    45 103
9   1     1   3   3 FALSE   0   0     0     0      0     0     0 103
10  1     1   4   1  TRUE   1   0     0     0      1     0    45 104
11  1     1   4   2 FALSE   1   1     1     1      0     1    25 104
12  1     1   4   3 FALSE   0   0     0     0      0     0     0 104
13  1     1   5   1 FALSE   1   1     1     0      1     0    35 105
14  1     1   5   2  TRUE   1   0     0     1      0     1    45 105
15  1     1   5   3 FALSE   0   0     0     0      0     0     0 105
16  1     1   6   1  TRUE   1   0     1     1      1     0    45 106
17  1     1   6   2 FALSE   1   1     0     0      0     1    25 106
18  1     1   6   3 FALSE   0   0     0     0      0     0     0 106
19  1     1   7   1 FALSE   1   1     0     1      1     1    25 107
20  1     1   7   2  TRUE   1   0     1     0      0     0    35 107
21  1     1   7   3 FALSE   0   0     0     0      0     0     0 107
22  1     1   8   1 FALSE   1   1     1     1      0     1    45 108
23  1     1   8   2  TRUE   1   0     0     0      1     0    25 108
24  1     1   8   3 FALSE   0   0     0     0      0     0     0 108
25  2     1   1   1 FALSE   1   1     1     1      0     0    45 201
26  2     1   1   2  TRUE   1   0     0     0      1     1    25 201
27  2     1   1   3 FALSE   0   0     0     0      0     0     0 201
28  2     1   2   1  TRUE   1   0     1     0      0     0    25 202
29  2     1   2   2 FALSE   1   1     0     1      1     1    35 202
30  2     1   2   3 FALSE   0   0     0     0      0     0     0 202
31  2     1   3   1 FALSE   1   0     0     1      0     0    35 203
32  2     1   3   2  TRUE   1   1     1     0      1     1    45 203
33  2     1   3   3 FALSE   0   0     0     0      0     0     0 203
34  2     1   4   1  TRUE   1   0     0     0      1     0    45 204
35  2     1   4   2 FALSE   1   1     1     1      0     1    25 204
```

Fig. 9.2 Converted data set using support.CEs (printed)

```
library(tidyverse)
df <- read.csv("d:\\R\\DCE1Impact_Formal.csv")
df <- df %>% select(-c(BLOCK))
dflong <- df %>% pivot_longer(!ID, names_to = "qs", values_to = "choice")
dflong <- dflong %>%
  mutate(
    p1 = case_when(choice == 1 ~ 1),
    p2 = case_when(choice == 2 ~ 1),
    p3 = case_when(choice == 3 ~ 1)
  )

dflong <- dflong %>% replace(is.na(.), 0)
dflong2 <- dflong %>% pivot_longer(cols = c("p1","p2","p3"),names_to = 'res')

res <- dflong2$value
```

The R code above reads a CSV file "DCE1Impact_Formal.csv" located in the "d:\R" directory into a data frame named *df*. It then removes a column named "BLOCK" from the data frame using the ***select()*** function from the **tidyverse** package. We then transformed the data frame *df* from a wide format to a long format using the ***pivot_longer()*** function, which creates a new column named *qs* that contains the original column names of the choice sets and a new column named *choice* that contains the values of the choice sets. We further transformed the dataset by creating three new columns (i.e., *p1*, *p2*, and *p3*) using the ***case_when()*** function that contains the value 1 if the corresponding value in the choice column is equal to 1, 2, or 3, respectively. Finally, we created a data frame called dflong2 by creating a new column named *res* that contains the original column names of the columns *p1*, *p2*, and *p3*, and a new column named *value* that contains the values of the corresponding cells in these columns. The values of the value *column* are then assigned to a variable named "res". Simply speaking, these R codes allow us to capture the DCE responses in a *binary vector* containing the choice data of all respondents (n = 237 in our sample project), as shown in Fig. 9.3.

After organizing the DCE responses, we also need to prepare a matrix of our experimental design that contains the same number of elements with the response vector (237 respondents × 3 alternatives × 8 choice sets = 5688 elements in total) for further processing. To prepare the design matrix, we can use the experiment design generated by the package **idefix** and repeat it 237 times using the following R code:

```
> res
   [1] 1 0 0 1 0 0 1 0 0 1 0 0 0 1 0 1 0 0 0 1 0 0 1 0 0 1 0 1 0 0 0 1 0 1 0 0 1 0 0 0 1 0 1 0 0 1 0 0 0 1 0
  [52] 1 0 0 1 0 0 1 0 0 0 1 0 0 1 0 0 0 1 0 0 1 1 0 0 1 0 0 1 0 0 1 0 0 0 1 0 1 0 0 0 1 0 0 1 0 0 1 0 1 0 0
 [103] 1 0 0 1 0 0 0 1 0 0 1 0 0 1 0 1 0 0 1 0 0 1 0 0 1 0 0 1 0 0 0 1 0 1 0 0 0 1 0 0 1 0 1 0 0 1 0 0 1 0 0
 [154] 1 0 0 0 1 0 1 0 0 0 1 0 0 1 0 1 0 0 0 1 0 0 0 1 1 0 0 1 0 0 0 0 1 1 0 0 0 1 0 0 1 0 0 1 0 1 0 0 1 0 0
 [205] 0 1 0 0 1 0 1 0 0 1 0 0 0 1 0 0 1 0 0 1 0 1 0 0 0 1 0 0 1 0 0 1 0 1 0 0 0 1 0 1 0 0 1 0 0 1 0 0 0 1 0
 [256] 0 1 0 0 1 0 0 1 0 0 1 0 1 0 0 1 0 0 1 0 0 0 1 0 0 1 0 0 1 0 1 0 0 1 0 0 1 0 0 1 0 0 1 0 0 0 1 0 1 0 0
 [307] 1 0 0 0 1 0 0 1 0 1 0 0 0 1 0 1 0 0 0 1 0 0 1 0 1 0 0 1 0 0 1 0 0 0 1 0 1 0 0 1 0 0 0 1 0 1 0 0 1 0 0
 [358] 1 0 0 1 0 0 1 0 0 1 0 0 1 0 0 0 1 0 1 0 0 0 1 0 0 1 0 0 1 0 1 0 0 1 0 0 1 0 0 0 1 0 1 0 0 0 1 0 0 0 1
 [409] 0 1 0 1 0 0 0 1 0 1 0 0 0 1 0 0 1 0 0 1 0 0 1 0 0 1 0 1 0 0 0 1 0 1 0 0 0 1 0 0 1 0 1 0 0 1 0 0 0 1 0
 [460] 1 0 0 0 1 0 1 0 0 1 0 0 0 1 0 1 0 0 0 1 0 0 1 0 1 0 0 0 1 0 1 0 0 0 1 0 0 1 0 1 0 0 0 1 0 0 1 0 1 0 0
 [511] 0 1 0 1 0 0 0 1 0 0 1 0 0 1 0 1 0 0 0 1 0 1 0 0 1 0 0 1 0 0 0 1 0 0 1 0 0 1 0 1 0 0 0 1 0 1 0 0 1 0 0
 [562] 1 0 0 0 1 0 0 1 0 0 1 0 1 0 0 0 1 0 1 0 0 0 1 0 1 0 0 1 0 0 0 1 0 1 0 0 0 1 0 1 0 0 1 0 0 1 0 0 1 0 0
 [613] 0 1 0 1 0 0 0 1 0 0 0 1 1 0 0 1 0 0 1 0 0 1 0 0 0 1 0 1 0 0 0 1 0 1 0 0 1 0 0 1 0 0 0 1 0 1 0 0 1 0 0
 [664] 0 1 0 0 1 0 1 0 0 1 0 0 1 0 0 1 0 0 1 0 0 0 1 0 1 0 0 0 1 0 1 0 0 0 1 0 1 0 0 1 0 0 1 0 0 0 1 0 0 1 0
 [715] 0 1 0 0 1 0 1 0 0 1 0 0 1 0 0 1 0 0 0 1 0 1 0 0 0 1 0 1 0 0 1 0 0 1 0 0 1 0 0 1 0 0 0 1 0 1 0 0 0 1 0
 [766] 0 0 1 0 1 0 1 0 0 1 0 0 1 0 0 0 1 0 0 1 0 0 1 0 1 0 0 1 0 0 1 0 0 1 0 0 1 0 0 0 1 0 0 1 0 0 1 0 1 0 0
 [817] 0 1 0 0 1 0 1 0 0 1 0 0 0 1 0 0 0 1 0 0 1 0 0 1 0 1 0 1 0 0 1 0 0 1 0 0 0 1 0 1 0 0 0 1 0 1 0 0 0 1 0
 [868] 1 0 0 1 0 0 1 0 0 0 1 0 1 0 0 0 1 0 1 0 0 0 1 0 1 0 0 0 1 0 1 0 0 0 1 0 0 1 0 0 1 0 0 1 0 1 0 0 0 1 0
 [919] 0 0 1 1 0 0 0 1 0 1 0 0 0 0 1 0 1 0 0 1 0 1 0 0 1 0 0 1 0 0 0 1 0 0 1 0 0 1 0 0 0 1 0 1 0 1 0 0 0 1 0
 [970] 1 0 0 1 0 0 0 1 0 0 1 0 1 0 0 0 1 0 1 0 0 1 0 0 1 0 0 0 1 0 0
 [ reached getOption("max.print") -- omitted 4688 entries ]
> |
```

Fig. 9.3 Printed DCE responses in a binary vector

library(idefix)
at.lvls <- c(2,2,2,2,2,3)
c.type <- c("D","D","D","D", "D", "C")
all <- Profiles(lvls = at.lvls, coding = c.type, c.lvls = con.lvls)
con.lvls <- list(c(3.5, 4.5, 6))
set.seed(123)
mu <- c(-0.6, 1.2, -0.2, 0.6, 0.4, -0.4,-0.6)
sigma <- diag(length(mu))
M <- MASS::mvrnorm(n = 10, mu = mu, Sigma = sigma)
pd <- list(matrix (M[,1],ncol = 1), M[,2:7])
D <- Modfed(cand.set = all, n.sets=8, n.alts = 3, no.choice = TRUE, alt.cte = c (0,0,1), par.draws = pd)
D
des1 <- as.matrix(D$design[, 2:7], ncol = 6) ##D is the object we created to save the experimental design

des <- des1[rep(1:nrow(des1),237),]

The printed output of the above code is shown in Fig. 9.4.

In this output, we created a design matrix with 8 choice sets and 237 × 3 alternatives. Each row of the matrix represents a profile or alternative, and each column represents an attribute with its corresponding levels.

Once we get both DCE responses and the experimental design ready, we can then use the function ***Datatrans()*** of the package **idefix** to convert our DCE data into various formats in order to use existing estimation packages in R (Traets et al., 2020). For instance, if we want to use the function ***rhierMnlRwMixture()*** from the package **bayesm** to estimate a hierarchical multinomial logit model, we run the following R code for data transformation:

```
> des
           Var12 Var22 Var32 Var42 Var52 Var6
set1.alt1      1     0     1     1     0  3.5
set1.alt2      1     1     1     1     0  3.5
no.choice      0     0     0     0     0  0.0
set2.alt1      0     0     1     1     0  3.5
set2.alt2      1     0     0     1     0  3.5
no.choice      0     0     0     0     0  0.0
set3.alt1      0     0     1     1     0  3.5
set3.alt2      1     1     0     1     0  3.5
no.choice      0     0     0     0     0  0.0
set4.alt1      1     0     1     1     0  4.5
set4.alt2      0     1     1     0     1  3.5
no.choice      0     0     0     0     0  0.0
set5.alt1      1     0     0     0     0  3.5
set5.alt2      1     0     1     1     1  3.5
no.choice      0     0     0     0     0  0.0
set6.alt1      1     0     1     0     1  3.5
set6.alt2      1     0     1     1     0  4.5
no.choice      0     0     0     0     0  0.0
set7.alt1      1     1     1     0     1  6.0
set7.alt2      1     0     1     0     0  3.5
no.choice      0     0     0     0     0  0.0
set8.alt1      1     1     1     0     0  3.5
set8.alt2      1     0     0     1     1  3.5
no.choice      0     0     0     0     0  0.0
set1.alt1      1     0     1     1     0  3.5
set1.alt2      1     1     1     1     0  3.5
no.choice      0     0     0     0     0  0.0
```

Fig. 9.4 A printed experimental design using idefix

Fig. 9.5 A converted DCE data using idefix

```
$lgtdata[[236]]
$lgtdata[[236]]$y
[1] 3 1 1 3 2 1 2 3

$lgtdata[[236]]$X
      var12 var22 var32 var42 var52 var6
 [1,]     1     0     1     1     0  3.5
 [2,]     1     1     1     1     0  3.5
 [3,]     0     0     0     0     0  0.0
 [4,]     0     0     1     1     0  3.5
 [5,]     1     0     0     1     0  3.5
 [6,]     0     0     0     0     0  0.0
 [7,]     0     0     1     1     0  3.5
 [8,]     1     1     0     1     0  3.5
 [9,]     0     0     0     0     0  0.0
[10,]     1     0     1     1     0  4.5
[11,]     0     1     1     0     1  3.5
[12,]     0     0     0     0     0  0.0
[13,]     1     0     0     0     0  3.5
[14,]     1     0     1     1     1  3.5
[15,]     0     0     0     0     0  0.0
[16,]     1     0     1     0     1  3.5
[17,]     1     0     1     1     0  4.5
[18,]     0     0     0     0     0  0.0
[19,]     1     1     1     0     1  6.0
[20,]     1     0     1     0     0  3.5
[21,]     0     0     0     0     0  0.0
[22,]     1     1     1     0     0  3.5
[23,]     1     0     0     1     1  3.5
[24,]     0     0     0     0     0  0.0
```

```
Datatrans(pkg = "bayesm", des = des, y= res, n.alts = 3, n.sets = 8, n.resp = 237, bin = TRUE)
## des is the experimental design, and res is DCE responses
```

The partial output of the function ***Datatrans()*** is shown in Fig. 9.5, which contains a list of data frames, each containing the transformed data for one respondent.

Using ***Datatrans()***, we can also convert the data set format into the required form using the package **Rchoice** for estimating various models such as Binary (logit and probit) and Ordered (logit and probit) models using simulated maximum likelihood (Sarrias, 2016):

```
Datatrans(pkg = "Rchoice", des = des, y= res, n.alts = 3, n.sets = 8, n.resp = 237, bin = TRUE)
```

The printed output is shown in Fig. 9.6 below:

9.3 Using R Packages for Model Estimation

After preparing and transforming the dataset, we can proceed with the model estimation. As mentioned, a wide range of R packages can be used to estimate different DCE models. In this section, we will briefly introduce how to use the function ***clogit()*** in the package **survival** (Therneau & Lumley, 2015), the ***mlogit()*** function in the package **mlogit** (Croissant, 2012), and the function ***apollo_estimate()*** in the package **apollo**

```
[1] "The dataset is ready to be used for Rchoice package"
   id.var    alt.names Var12 Var22 Var32 Var42 Var52 Var6 Choice
1       1 alternative.1     1     0     1     1     0  3.5      1
2       1 alternative.2     1     1     1     1     0  3.5      0
3       1 alternative.3     0     0     0     0     0    0      0
4       1 alternative.1     0     0     1     1     0  3.5      1
5       1 alternative.2     1     0     0     1     0  3.5      0
6       1 alternative.3     0     0     0     0     0    0      0
7       1 alternative.1     0     0     1     1     0  3.5      1
8       1 alternative.2     1     1     0     1     0  3.5      0
9       1 alternative.3     0     0     0     0     0    0      0
10      1 alternative.1     1     0     1     1     0  4.5      1
11      1 alternative.2     0     1     1     0     1  3.5      0
12      1 alternative.3     0     0     0     0     0    0      0
13      1 alternative.1     1     0     0     0     0  3.5      0
14      1 alternative.2     1     0     1     1     1  3.5      1
15      1 alternative.3     0     0     0     0     0    0      0
16      1 alternative.1     1     0     1     0     1  3.5      1
17      1 alternative.2     1     0     1     1     0  4.5      0
18      1 alternative.3     0     0     0     0     0    0      0
19      1 alternative.1     1     1     1     0     1    6      0
20      1 alternative.2     1     0     1     0     0  3.5      1
21      1 alternative.3     0     0     0     0     0    0      0
22      1 alternative.1     1     1     1     0     0  3.5      0
23      1 alternative.2     1     0     0     1     1  3.5      1
24      1 alternative.3     0     0     0     0     0    0      0
25      2 alternative.1     1     0     1     1     0  3.5      0
26      2 alternative.2     1     1     1     1     0  3.5      1
27      2 alternative.3     0     0     0     0     0    0      0
28      2 alternative.1     0     0     1     1     0  3.5      1
29      2 alternative.2     1     0     0     1     0  3.5      0
30      2 alternative.3     0     0     0     0     0    0      0
31      2 alternative.1     0     0     1     1     0  3.5      0
32      2 alternative.2     1     1     0     1     0  3.5      1
```

Fig. 9.6 A screenshot of converted DCE data using idefix (Cont)

(Hess & Palma, 2019) for model estimation. We will estimate the MNL model, as it is one of the most commonly used models in DCE studies (Anas, 1983; Soekhai et al., 2019).

We used the package **support.CEs** to generate the experimental design for our demonstrative project (i.e., that assesses consumers' preferences for alternative meat products), and thus the design matrix that we need for model estimation can be created using the following R code:

```
desmat2 <- make.design.matrix(choice.experiment.design = unlabeled, optout = TRUE,
                    categorical.attributes  =  c("Additives",  "Nutrition",  "Convenience",
"CarbonFootprint", "VeganLabel"),
                    continuous.attributes = c("Price"),
                    unlabeled = TRUE)
```

##unlabeled refers to our experimental design generated using the function ***rotation.design()***. More details can be found in section 7.5.1 in Chapter 7.

```
> desmat2
   BLOCK QES ALT ASC Yes Forty Ready Twenty Yes.2 Price
1      1   1   1   1   1     1     1      0     0    45
2      1   1   2   1   0     0     0      1     1    25
3      1   1   3   0   0     0     0      0     0     0
4      1   2   1   1   0     1     0      0     0    25
5      1   2   2   1   1     0     1      1     1    35
6      1   2   3   0   0     0     0      0     0     0
7      1   3   1   1   0     0     1      0     0    35
8      1   3   2   1   1     1     0      1     1    45
9      1   3   3   0   0     0     0      0     0     0
10     1   4   1   1   0     0     0      1     0    45
11     1   4   2   1   1     1     1      0     1    25
12     1   4   3   0   0     0     0      0     0     0
13     1   5   1   1   1     1     0      1     0    35
14     1   5   2   1   0     0     1      0     1    45
15     1   5   3   0   0     0     0      0     0     0
16     1   6   1   1   0     1     1      1     0    45
17     1   6   2   1   1     0     0      0     1    25
18     1   6   3   0   0     0     0      0     0     0
19     1   7   1   1   1     0     1      1     1    25
20     1   7   2   1   0     1     0      0     0    35
21     1   7   3   0   0     0     0      0     0     0
22     1   8   1   1   1     1     1      0     1    45
23     1   8   2   1   0     0     0      1     0    25
24     1   8   3   0   0     0     0      0     0     0
25     2   1   1   1   1     0     1      0     0    25
26     2   1   2   1   0     1     0      1     1    35
27     2   1   3   0   0     0     0      0     0     0
```

Fig. 9.7 Printed design matrix using support.CEs

The printed design matrix is shown in Fig. 9.7.

9.3.1 *Logit Analysis Using clogit Function*

After the data set is also converted and ready for use and saved in the object *dataset2* (see Sect. 9.2), we can then use the function ***clogit()*** in the package **survival** for model estimation:

```
library(survival)
clogout1 <- clogit(RES ~ ASC + Yes + Forty + Ready + Price + Twenty + Yes.2 + strata(STR),
data = dataset2)
clogout1
```

In the R code above, "Yes", "Forty", "Ready", "Twenty", and "Yes.2" correspond to the names of the attribute levels of the categorical attributes "Additives", "Nutrition", "Convenience", "CarbonFootprint", and "VeganLabel". The first attribute levels were omitted in the dummy coding. In the formula provided to ***clogit()***, each attribute level is included as a predictor variable along with the **ASC** (alternative-specific constant) and "Price" continuous attribute.

```
Call:
clogit(RES ~ ASC + Yes + Forty + Ready + Price + Twenty + Yes.2 +
    strata(STR), data = dataset2)

           coef exp(coef) se(coef)      z        p
ASC     2.95717  19.24335  0.17636 16.768  < 2e-16
Yes    -0.11661   0.88993  0.04750 -2.455  0.01409
Forty   0.13784   1.14779  0.04750  2.902  0.00371
Ready  -0.09596   0.90850  0.04749 -2.021  0.04332
Price  -0.11074   0.89517  0.02655 -4.171 3.04e-05
Twenty  0.14410   1.15501  0.04750  3.034  0.00242
Yes.2  -0.06190   0.93997  0.04750 -1.303  0.19245

Likelihood ratio test=1042  on 7 df, p=< 2.2e-16
n= 5688, number of events= 1896
```

Fig. 9.8 Printed model estimation results using survival

The **coefficients of the MNL model** will be estimated by **maximum likelihood estimation** (Hoyos, 2010). They will be printed on the R console (see Fig. 9.8) after calling the ***clogit()*** function based on the model specification and the abovementioned data set.

The output shows each attribute's coefficients, standard errors, z-values, and p-values. It also provides the likelihood ratio test statistics, which compares the fit of two models, one nested within the other (i.e., it tells us whether our full model with all attributes better fits the data than the null model without attributes).

Besides calculating the main effects, we can also estimate the interaction effects by adding additional variables to the analysis. For instance, we want to examine whether the demographic backgrounds of respondents will influence their preference for alternative meat products, we can add new columns, such as "**Gender**" (*1 = Male; 0 = Female*), "**Age**" (*1 = 18 – 25; 2 = 26–35; 3 = 36–45; 4 = 46–55; 5 = 56–65; 6 = 66–75; 7 = 76 or above*), and "**Education**" (*1 = Primary School or Below; 2 = Secondary School; 3 = Diploma; 4 = Associate Degree/High Diploma; 5 = Bachelor's Degree; 6 = Postgraduate or above*) in the dataset (see Fig. 9.9).

Then, we can run the following R codes to compute the interaction effects between the attribute levels and demographic variables (Gender, Age, and Education):

```
library(survival)
clogout2 <- clogit(RES ~ ASC + Yes + Forty + Ready + Price + Twenty + Yes.2
            + Yes:Gender + Forty:Gender + Ready:Gender + Price:Gender + Twenty:Gender +
Yes.2:Gender
            + Yes:Age + Forty:Age + Ready:Age + Price:Age + Twenty:Age + Yes.2:Age
            + Yes:Education + Forty:Education + Ready:Education + Price:Education +
Twenty:Education + Yes.2:Education
            + strata(STR), data = dataset2)
Clogout2
```

	A	B	C	D	E	F	G	H	I	J	K	L	M
1	ID	BLOCK	q1	q2	q3	q4	q5	q6	q7	q8	Gender	Age	Education
2	1	1	3	3	3	3	1	3	3	3	1	4	5
3	2	1	3	3	3	3	3	3	3	3	1	3	6
4	3	1	3	3	3	3	3	3	3	3	0	2	6
5	4	1	1	1	2	2	1	2	1	1	0	3	3
6	5	1	1	1	2	2	1	2	1	1	0	2	5
7	6	1	1	1	1	2	1	2	3	1	1	3	5
8	7	1	1	1	2	2	1	2	2	2	0	2	6
9	8	1	1	1	2	1	1	3	1	1	1	5	4
10	9	1	1	2	2	1	1	1	2	2	1	3	6
11	10	1	1	1	2	2	1	2	1	1	1	3	3
12	11	1	1	3	1	2	1	3	3	3	0	3	5
13	12	1	1	1	1	2	2	2	1	1	0	2	5
14	13	1	1	1	2	2	1	2	1	1	0	3	5
15	14	1	1	1	2	1	1	1	1	1	0	2	4
16	15	1	1	1	2	1	2	1	1	2	1	2	5
17	16	1	1	1	1	2	1	2	1	1	0	3	3
18	17	1	1	3	2	2	1	2	3	3	0	2	5
19	18	1	1	3	2	3	1	3	2	2	0	2	4
20	19	1	1	1	2	1	1	2	1	1	1	3	5
21	20	1	1	1	1	2	1	2	1	1	1	5	6
22	21	1	1	1	2	3	1	3	1	1	1	3	5
23	22	1	1	1	1	2	1	2	1	1	0	4	5

Fig. 9.9 Adding demographic variables

In the updated R codes, we added additional interaction terms. For example, "*Yes:Gender*" is an interaction term between the attribute level "Yes" of the "Additives" attribute and the demographic variable 'Gender'. This term allows for different preferences for the attribute level 'Yes' based on the gender of the respondent. Similarly, the other interaction terms allow for differential preferences for the attribute levels based on the respondent's age and education level. Importantly, including these interaction terms in the model allows for a more nuanced analysis of the data, as it accounts for potential differences in preferences across demographic groups, unpacking the preference heterogeneity in the respondents. The printed estimation results are shown in Fig. 9.10 (we will discuss how to interpret these results in Chap. 10):

9.3.2 *Generating Goodness of Fit Results Using gofm() Function*

The function ***clogit()*** returns estimated coefficients about a DCE model and the Likelihood ratio test. However, it **does not measure the goodness of fit.** Unlike the linear regression model, there is no single measure for the goodness of fit in discrete choice models (Kjær, 2005). The most commonly used measure of **goodness of fit is the McFadden R^2,** also known as pseudo R^2.

The R^2 can be calculated using the log-likelihood values of the logit models, which is an indicator of the relative explanatory power of a model (Hauber et al., 2016). It is defined as follows (where LL is the log-likelihood value):

```
Call:
clogit(RES ~ ASC + Yes + Forty + Ready + Price + Twenty + Yes.2 +
    Yes:Gender + Forty:Gender + Ready:Gender + Price:Gender +
    Twenty:Gender + Yes.2:Gender + Yes:Age + Forty:Age + Ready:Age +
    Price:Age + Twenty:Age + Yes.2:Age + Yes:Education + Forty:Education +
    Ready:Education + Price:Education + Twenty:Education + Yes.2:Education +
    strata(STR), data = dataset2)

                     coef exp(coef)  se(coef)      z       p
ASC              2.967444 19.442164  0.176827 16.782 <2e-16
Yes             -0.373519  0.688308  0.296379 -1.260 0.2076
Forty            0.089829  1.093987  0.296749  0.303 0.7621
Ready           -0.348000  0.706099  0.296245 -1.175 0.2401
Price           -0.001549  0.998452  0.121753 -0.013 0.9898
Twenty           0.257699  1.293949  0.296072  0.870 0.3841
Yes.2            0.072947  1.075674  0.295848  0.247 0.8052
Yes:Gender       0.041855  1.042743  0.097546  0.429 0.6679
Forty:Gender     0.121615  1.129319  0.097486  1.248 0.2122
Ready:Gender     0.039231  1.040010  0.097295  0.403 0.6868
Price:Gender    -0.008712  0.991326  0.039672 -0.220 0.8262
Twenty:Gender    0.041481  1.042354  0.097586  0.425 0.6708
Yes.2:Gender    -0.100712  0.904193  0.097485 -1.033 0.3016
Yes:Age          0.110661  1.117017  0.044631  2.479 0.0132
Forty:Age       -0.018852  0.981325  0.044437 -0.424 0.6714
Ready:Age        0.008380  1.008415  0.044442  0.189 0.8504
Price:Age       -0.018474  0.981695  0.018080 -1.022 0.3069
Twenty:Age       0.006308  1.006328  0.044601  0.141 0.8875
Yes.2:Age       -0.027808  0.972575  0.044632 -0.623 0.5332
Yes:Education   -0.016542  0.983594  0.051962 -0.318 0.7502
Forty:Education  0.010189  1.010241  0.051773  0.197 0.8440
Ready:Education  0.044366  1.045364  0.051799  0.856 0.3917
Price:Education -0.011610  0.988457  0.021191 -0.548 0.5838
Twenty:Education -0.030873  0.969599  0.051920 -0.595 0.5521
Yes.2:Education -0.002600  0.997403  0.051837 -0.050 0.9600

Likelihood ratio test=1056  on 25 df, p=< 2.2e-16
n= 5688, number of events= 1896
```

Fig. 9.10 Printed estimation results with interaction effects

$$\text{McFadden } R^2 = 1 - \frac{\text{LL of model}}{\text{LL model without predictors}}$$

When the value of the McFadden R^2 equals 1, it means that the respondent's choice can be perfectly predicted. Nonetheless, the measure can never reach 1. Other goodness of fit statistics exist, such as the "*percent correctly predicted*". However, this statistic was criticized for providing less information about the model than the McFadden R^2 (Kjær, 2005).

To calculate the McFadden R^2, we can use the function ***gofm()*** in the package **support.CEs,** which provides R^2 and the R^2 adjusted by the number of estimated coefficients (Aizaki, 2012). After we execute the ***clogit()*** function in the package survival, we can then run the following code for calculating the goodness of fit of the estimated model:

gofm(clogout1)

```
> gofm(clogout1)

Rho-squared = 0.2500973
Adjusted rho-squared = 0.2467367
Akaike information criterion (AIC) = 3138.048
Bayesian information criterion (BIC) = 3176.88
Number of coefficients = 7
Log likelihood at start = -2082.969
Log likelihood at convergence = -1562.024
```

Fig. 9.11 Results of the function *gofm()*

The returned results are shown in Fig. 9.11:

9.3.3 *Estimating the Willingness to Pay of Non-monetary Attributes*

Calculating how much respondents are willing to forgo one attribute in favor of another attribute is another useful analysis that can be done in DCE. This analysis is also known as attribute trade-off analysis or **marginal willingness to pay (MWTP)** analysis. The function ***mwtp()*** in the package **support.CEs** can help us calculate the MWTP values of non-monetary variables. MWTP values are monetary value estimates by dividing the coefficient of each non-monetary attribute by the coefficient of the monetary attribute. For example, if the coefficient of the "vegan label" attribute in a DCE model is 0.5, and the coefficient of the "price" attribute is -1, then the MWTP for the "vegan label" attribute would be $0.5/{-1} = -0.5$, indicating that respondents are willing to pay an additional \$0.50 for the product with the vegan label compared to a product without one.

Particularly, MWTP can be defined using the following formula where $\boldsymbol{\beta}_{nm}$ refers to coefficients of non-monetary attributes and $\boldsymbol{\beta}_{m}$ refers to the coefficient of the monetary attribute (Krinsky & Robb, 1986):

$$\mathrm{WTP} = \frac{\beta_{nm}}{\beta_m}$$

We can execute the following R code to estimate the marginal WTP for each non-monetary attribute by specifying the monetary attribute and the non-monetary attributes as well as the confidence level using the function ***mwtp()*** (Aizaki, 2012):

```
mwtp(output = clogout1, monetary.variables = c("Price"),
  non-monetary.variables = c("Yes", "Forty", "Ready", "Twenty", "Yes.2"),
  confidence.level = c(0.9), seed = 987)
```

This ***mwtp()*** function takes several arguments:

- **output**: the output from the ***clogit()*** function, which contains the coefficients of the model.
- **monetary.variables**: a character vector of the names of the variables that are considered as monetary attributes (in this case, just "Price").
- **non-monetary.variables**: a character vector of the names of the variables that are considered as non-monetary attributes.
- **confidence.level**: the level of confidence desired for the confidence intervals (in this case, 90%).
- **seed**: an optional seed for the random number generator.

After we run the R code, he MWTP values will be printed on the R console as follows (Fig. 9.12):

The output above shows the MWTP values for each non-monetary attribute, along with their 90% confidence intervals. For example, the results demonstrate that the respondents are willing to pay an extra $1.05 to buy an alternative meat product without additives.

Other R packages can also be used for model estimation. For example, we can use the ***mlogit()*** function in the package **mlogit** to estimate the MNL model after we convert the data set into the required format by using the *Datatrans* function:

```
install.packages("mlogit")
library(mlogit)
mlogit <- Datatrans(pkg = "mlogit", des = des, y= res, n.alts = 3, n.sets = 8, n.resp = 237, bin =
TRUE)
m1 <- mlogit(Choice ~ 0 + Var12 + Var22 + Var32 + Var42 + Var52 + Var6, data = mlogit)
summary(m1)
```

The estimated coefficients will be printed in the R console, as shown in Fig. 9.13.

```
> mwtp(output = clogout1, monetary.variables = c("Price"),
+       nonmonetary.variables = c("Yes", "Forty", "Ready", "Twenty", "Yes.2"),
+       confidence.level = c(0.9), seed = 987)

           MWTP      5%     95%
Yes     -1.0530 -2.1065 -0.3335
Forty    1.2448  0.5141  2.4258
Ready   -0.8666 -1.8906 -0.1565
Twenty   1.3013  0.5568  2.4875
Yes.2   -0.5590 -1.4380  0.1543

method = Krinsky and Robb
```

Fig. 9.12 Returned results of the function *mwtp*

```
Call:
mlogit(formula = choice ~ 0 + Var12 + Var22 + Var32 + Var42 +
    Var52 + Var6, data = cbc.logit, method = "nr")

Frequencies of alternatives:choice
alternative.1 alternative.2 alternative.3
     0.489451      0.433544      0.077004

nr method
5 iterations, 0h:0m:0s
g'(-H)^-1g = 4.35E-08
gradient close to zero

Coefficients :
          Estimate Std. Error z-value  Pr(>|z|)
Var121   -0.210050   0.108045 -1.9441 0.0518831 .
Var221   -0.534912   0.081932 -6.5288 6.632e-11 ***
Var321    0.477384   0.088442  5.3977 6.750e-08 ***
Var421    0.422030   0.130276  3.2395 0.0011974 **
Var521   -0.114388   0.127087 -0.9001 0.3680771
Var63.5   1.558858   0.169775  9.1819 < 2.2e-16 ***
Var64.5   1.062978   0.317729  3.3455 0.0008212 ***
Var66     1.830853   0.273531  6.6934 2.180e-11 ***
---
Signif. codes:  0 '***' 0.001 '**' 0.01 '*' 0.05 '.' 0.1 ' ' 1

Log-Likelihood: -1642.7
```

Fig. 9.13 Printed model estimation results using *mlogit()*

9.3.4 Model Estimation Using Other R Packages

Another popular package is **apollo**, which also allows choice model estimation in the R platform (Hess & Palma, 2019). This package can be used together with **support.CEs,** where the latter helps with preparing the data set into a suitable format. Hess and Palma (2019) developed the package **apollo** package and provides a flexible tool for model estimation of DCE studies. Besides supporting the estimation of the classical conditional logit models, it also allows users to write their own model functions.[3]

To start with, we will first need to reshape the data set using the function *reshape* in the **stats** package and create a variable indicating the responses to DCE questions:

```
dataset1r <- reshape(
  dataset2, timevar = "ALT", idvar= "STR",
  v.names = c("RES","ASC","Yes","Forty","Ready","Twenty","Yes.2","Price"),
  direction = "wide")
dataset1r$choice <- dataset1r$RES.1 * 1 + dataset1r$RES.2 * 2 + dataset1r$RES.3 * 3
head(dataset1r)
##dataset2 is the dataset generated using the make.dataset() function in support.CEs
```

[3] More information about the package and its applications for estimating different models can be found in http://www.ApolloChoiceModelling.com.

```
> dataset1r$choice <- dataset1r$RES.1 * 1 + dataset1r$RES.2 * 2 + dataset1r$RES.3 * 3
> head(dataset1r)
   ID BLOCK QES STR RES.1 ASC.1 Yes.1 Forty.1 Ready.1 Twenty.1 Yes.2.1 Price.1 RES.2 ASC.2 Yes.2 Forty.2
1   1     1   1 101  TRUE     1     1       1       1        0       0      45 FALSE     1     0       0
4   1     1   2 102  TRUE     1     0       1       0        0       0      25 FALSE     1     1       0
7   1     1   3 103  TRUE     1     0       0       1        0       0      35 FALSE     1     1       1
10  1     1   4 104  TRUE     1     0       0       0        1       0      45 FALSE     1     1       1
13  1     1   5 105 FALSE     1     1       1       0        1       0      35  TRUE     1     0       0
16  1     1   6 106  TRUE     1     0       1       1        1       0      45 FALSE     1     1       0
   Ready.2 Twenty.2 Yes.2.2 Price.2 RES.3 ASC.3 Yes.3 Forty.3 Ready.3 Twenty.3 Yes.2.3 Price.3 choice
1        0        1       1      25 FALSE     0     0       0       0        0       0       0      1
4        1        1       1      35 FALSE     0     0       0       0        0       0       0      1
7        0        1       1      45 FALSE     0     0       0       0        0       0       0      1
10       1        0       1      25 FALSE     0     0       0       0        0       0       0      1
13       1        0       1      45 FALSE     0     0       0       0        0       0       0      2
16       0        0       1      25 FALSE     0     0       0       0        0       0       0      1
```

Fig. 9.14 Reshaped dataset for apollo

The reshaped dataset that is ready to be used in the package **apollo** is shown in Fig. 9.14:

After preparing the dataset, the next step is to initialize the package **apollo** and prepare the inputs by specifying the model name and a brief description using the following commands:

```
library(apollo)
apollo_initialise()
apollo_control <- list(
  modelName = "CL_apollo",
  modelDescr = "Consumers' preferences for alternative meat products",
  indivID = "ID")
```

Next, we can load our converted dataset to the database in **apollo** and define the parameter names and their starting values accordingly:

```
database <- dataset1
apollo_beta <- c(
  asc     = 0,
  b_Yes   = 0,
  b_Forty  = 0,
  b_Ready  = 0,
  b_Twenty  = 0,
  b_Yes.2  = 0,
  b_Price = 0
)
apollo_fixed <- c()
apollo_inputs <- apollo_validateInputs()
```

The final preparation step is to define the function ***apollo_probabilities()*** by using the following R code:

```
apollo_probabilities <- function(apollo_beta, apollo_inputs, functionality = "estimate"){
  apollo_attach(apollo_beta, apollo_inputs)
  on.exit(apollo_detach(apollo_beta, apollo_inputs))

  P <- list()
  V <- list()

  V[['alt1']] = asc + b_Yes*Yes.1 + b_Forty*Forty.1 + b_Ready*Ready.1 + b_Twenty*Twenty.1
+ b_Yes.2*Yes.2.1 +
   b_Price*Price.1
  V[['alt2']] = asc + b_Yes*Yes.2 + b_Forty*Forty.2 + b_Ready*Ready.2 + b_Twenty*Twenty.2
+ b_Yes.2*Yes.2.2 +
   b_Price*Price.2
  V[['alt3']] = 0

  mnl_settings <- list(
   alternatives = c(alt1 = 1, alt2 = 2, alt3 = 3),
   avail        = list(alt1 = 1, alt2 = 1, alt3 = 1),
   choiceVar    = choice,
   V            = V
  )
  P[['model']] <- apollo_mnl(mnl_settings, functionality)
  P <- apollo_panelProd(P, apollo_inputs, functionality)
  P <- apollo_prepareProb(P, apollo_inputs, functionality)

  return(P)
}
```

The function ***apollo_probabilities()*** takes as inputs the coefficients of the MNL model (b_Yes, b_Forty, b_Ready, b_Twenty, b_Yes.2, and b_Price), the intercept (asc), the data on the alternatives and the choice made by the individuals. The function then calculates the utility of each alternative using the coefficients and the data and estimates the probabilities of choosing each alternative using the MNL model.

Finally, the MNL model can be estimated using the function ***apollo_estimate()***:

```
model <- apollo_estimate(apollo_beta, apollo_fixed, apollo_probabilities, apollo_inputs)
```

The summary of the results can be printed on the R console (see Fig. 9.15) by executing the following code:

```
apollo_modelOutput(model, list(printPVal = TRUE))
```

The MWTP values for non-monetary attributes (e.g., additives, vegan label, the format of alternative meat products) can be calculated using the function ***apollo_deltaMethod()*** using the following R code:

```
> apollo_modelOutput(model, list(printPVal = TRUE))
Model run by Echo using Apollo 0.2.8 on R 4.2.0 for Windows.
www.ApolloChoiceModelling.com

Model name                                  : CL_apollo
Model description                           : Consumers' preferences for alternative meat products
Model run at                                : 2022-12-19 23:37:56
Estimation method                           : bfgs
Model diagnosis                             : successful convergence
Number of individuals                       : 237
Number of rows in database                  : 1896
Number of modelled outcomes                 : 1896

Number of cores used                        :  1
Model without mixing

LL(start)                                   : -2082.97
LL at equal shares, LL(0)                   : -2082.97
LL at observed shares, LL(C)                : -1724.35
LL(final)                                   : -1645.74
Rho-squared vs equal shares                 :  0.2099
Adj.Rho-squared vs equal shares             :  0.2065
Rho-squared vs observed shares              :  0.0456
Adj.Rho-squared vs observed shares          :  0.0415
AIC                                         :  3305.47
BIC                                         :  3344.3

Estimated parameters                        :  7
Time taken (hh:mm:ss)                       :  00:00:0.38
     pre-estimation                         :  00:00:0.14
     estimation                             :  00:00:0.12
     post-estimation                        :  00:00:0.12
Iterations                                  :  17
Min abs eigenvalue of Hessian               :  43.15968

Unconstrained optimisation.

These outputs have had the scaling used in estimation applied to them.
Estimates:
           Estimate        s.e.   t.rat.(0)  p(1-sided)    Rob.s.e. Rob.t.rat.(0)  p(1-sided)
asc        2.430115    0.149772      16.225     0.00000    0.180474        13.465     0.00000
b_Yes     -0.304399    0.065240      -4.666   1.537e-06    0.057567        -5.288   6.193e-08
b_Forty   -0.283136    0.064260      -4.406   5.262e-06    0.043820        -6.461   5.188e-11
b_Ready    0.124197    0.060742       2.045     0.02044    0.075508         1.645     0.05000
b_Twenty  -0.335973    0.055096      -6.098   5.374e-10    0.055321        -6.073   6.270e-10
b_Yes.2   -0.305229    0.067213      -4.541   2.796e-06    0.062363        -4.894   4.930e-07
b_Price   -0.003900    0.003657      -1.066     0.14316    0.003718        -1.049     0.14713
```

Fig. 9.15 Printed model estimation results using apollo

```
for (i in c("b_Yes", "b_Forty", "b_Ready", "b_Twenty", "b_Yes.2")) {
  deltaMethod_settings <- list(operation = "ratio", parName1 = i, parName2 = "b_Price",
                               multPar1 = -1)
  apollo_deltaMethod(model, deltaMethod_settings)
}
```

The results will then be shown on the R console, as shown in Fig. 9.16 below.

It is important to note that many R packages also support the estimation of models other than the classical MNL (e.g., the package **apollo** provides functions for many models and allows users to add new structures to customize their model estimation). Users need to decide which model to use for their DCE data estimation before choosing appropriate R packages for data preparation and analysis.

The next chapter will briefly introduce how we can report DCE findings based on the model estimation results.

```
Running Delta method computations
                                                  Value Robust s.e. Rob t-ratio (0)
Ratio of b_Yes (multiplied by -1) and b_Price:   -78.06       79.86          -0.9774

Running Delta method computations
                                                    Value Robust s.e. Rob t-ratio (0)
Ratio of b_Forty (multiplied by -1) and b_Price:   -72.6        72.08          -1.007

Running Delta method computations
                                                    Value Robust s.e. Rob t-ratio (0)
Ratio of b_Ready (multiplied by -1) and b_Price:   31.85        29.2            1.091

Running Delta method computations
                                                     Value Robust s.e. Rob t-ratio (0)
Ratio of b_Twenty (multiplied by -1) and b_Price:   -86.15       89.84          -0.959

Running Delta method computations
                                                    Value Robust s.e. Rob t-ratio (0)
Ratio of b_Yes.2 (multiplied by -1) and b_Price:   -78.27       74.16          -1.055
```

Fig. 9.16 Printed results of WTP values using apollo

References

Aizaki, H. (2012). Basic functions for supporting an implementation of choice experiments in R. *Journal of Statistical Software, 50*, 1–24.

Anas, A. (1983). Discrete choice theory, information theory and the multinomial logit and gravity models. *Transportation Research Part B: Methodological, 17*(1), 13–23.

Bech, M., & Gyrd-Hansen, D. (2005). Effects coding in discrete choice experiments. *Health Economics, 14*(10), 1079–1083.

Bliemer, M. C., Rose, J. M., & Hensher, D. A. (2009). Efficient stated choice experiments for estimating nested logit models. *Transportation Research Part B: Methodological, 43*(1), 19–35.

Campbell, D., & Erdem, S. (2019). Including opt-out options in discrete choice experiments: Issues to consider. *The Patient-Patient-Centered Outcomes Research, 12*(1), 1–14.

Croissant, Y. (2012). Estimation of multinomial logit models in R: The mlogit Packages. *R package version 0.2-2*. http://cran.r-project.org/web/packages/mlogit/vignettes/mlogit.pdf

De Bekker-Grob, E. W., Ryan, M., & Gerard, K. (2012). Discrete choice experiments in health economics: A review of the literature. *Health Economics, 21*(2), 145–172.

Hauber, A. B., González, J. M., Groothuis-Oudshoorn, C. G., Prior, T., Marshall, D. A., Cunningham, C., … & Bridges, J. F. (2016). Statistical methods for the analysis of discrete choice experiments: A report of the ISPOR conjoint analysis good research practices task force. *Value in health, 19*(4), 300–315.

Hensher, D. A., Rose, J. M., Rose, J. M., & Greene, W. H. (2005). *Applied choice analysis: A primer*. Cambridge University Press.

Hess, S., & Palma, D. (2019). Apollo: A flexible, powerful and customizable freeware package for choice model estimation and application. *Journal of Choice Modelling, 32*, 100170.

Hoyos, D. (2010). The state of the art of environmental valuation with discrete choice experiments. *Ecological Economics, 69*(8), 1595–1603.

Kjaer, T. (2005). *A review of the discrete choice experiment-with emphasis on its application in health care*. Syddansk Universitet.

Krinsky, I., & Robb, A. L. (1986). On approximating the statistical properties of elasticities. *The review of economics and statistics*, 715–719.

Lancsar, E., & Louviere, J. (2008). Conducting discrete choice experiments to inform healthcare decision making. *PharmacoEconomics, 26*(8), 661–677.

Louviere, J. J., Hensher, D. A., & Swait, J. D. (2000). *Stated choice methods: Analysis and applications*. Cambridge University Press.

Mangham, L. J., Hanson, K., & McPake, B. (2009). How to do (or not to do) ... Designing a discrete choice experiment for application in a low-income country. *Health Policy and Planning, 24*(2), 151–158.

Mazzanti, M. (2003). Discrete choice models and valuation experiments. *Journal of Economic Studies, 30*(6), 584–604.

McFadden, D. (1974). The measurement of urban travel demand. *Journal of Public Economics, 3*(4), 303–328.

McFadden, D., & Train, K. (2000). Mixed MNL models for discrete response. *Journal of Applied Econometrics, 15*(5), 447–470.

Roux, L., Ubach, C., Donaldson, C., & Ryan, M. (2004). Valuing the benefits of weight loss programs: An application of the discrete choice experiment. *Obesity Research, 12*(8), 1342–1351.

Sarrias, M. (2016). Discrete choice models with random parameters in R: The Rchoice package. *Journal of Statistical Software, 74*, 1–31.

Soekhai, V., de Bekker-Grob, E. W., Ellis, A. R., & Vass, C. M. (2019). Discrete choice experiments in health economics: Past, present and future. *PharmacoEconomics, 37*(2), 201–226.

Steckel, J. H., & Vanhonacker, W. R. (1988). A heterogeneous conditional logit model of choice. *Journal of Business & Economic Statistics, 6*(3), 391–398.

Therneau, T. M., & Lumley, T. (2015). Package 'survival.' *R Top Doc, 128*(10), 28–33.

Traets, F., Sanchez, D. G., & Vandebroek, M. (2020). Generating optimal designs for discrete choice experiments in R: The Idefix package. *Journal of Statistical Software, 96*, 1–41.

Train, K. E. (2009). *Discrete choice methods with simulation*. Cambridge University Press.

Chapter 10
Visualizing and Reporting DCE Data Using R

A well-designed and clear visual representation of the data is essential to communicate the results to a wide range of stakeholders. In addition, visualizing the data can also help researchers explore patterns and relationships within the data, identify outliers, and inform further the analysis. In this chapter, we will first describe different methods for visualizing DCE data in R and then outline the steps for reporting the results.

10.1 Visualizing Data Using R

Several packages in R can be used to visualize DCE data, including **ggplot2**, **lattice**, and **plotly**. In particular, the package **ggplot2** (Wickham, 2011) is a popular and versatile package for visualizing data in R and provides a variety of plots such as bar plots, histograms, scatterplots, and density plots. The package **lattice** (Sarkar, 2008) is also useful for visualizing multivariate data, including data from DCE studies. Finally, **plotly** (Sievert, 2020) is a package that allows for the creation of interactive plots and helps present data engagingly through various visualizations, including bar charts, line plots, scatter plots, and histograms. As mentioned in previous chapters, these packages can be installed in R using the ***install.packages()*** function and load them into the R environment using the ***library()*** function. Alternatively, we can rely on the base functions in R, such as ***barplot()*** and ***hist()***, for some simple visualizations.

The visualization of DCE data can be divided into two parts. The first is to visualize graphic information of DCE respondents, and the second is to visualize the core choice data and DCE results. Bar plots, histograms, and box plots are commonly used graphs to visualize demographic information. We will use our sample DCE data (see Fig. 10.1) to demonstrate how to visualize demographic data in R.

L. Shang and Y. Chandra, *Discrete Choice Experiments Using R*,
https://doi.org/10.1007/978-981-99-4562-7_10

	A	B	C	D	E	F	G	H	I	S	T
1	Gender	Age	Religion	MarriageS	Country	Education	Employm	IncomeLe	DietHabit	Environmental Concern	Behav
2	1	4	5	3	4	5	3	3	2	2.8	2.333333
3	1	3	5	3	4	6	3	2	6	2.8	4.666667
4	0	2	1	2	4	6	2	4	1	2.6	3.333333
5	0	3	1	1	3	3	2	2	4	4.6	4.333333
6	0	2	5	1	4	5	2	6	1	3.6	4
7	1	3	1	3	4	5	2	6	2	4	3.666667
8	0	2	2	2	4	6	2	6	1	2.8	3
9	1	5	5	1	4	4	3	3	1	5	5
10	1	3	5	2	4	6	2	6	1	4	4.333333
11	1	3	5	2	4	3	3	4	1	1.6	2.666667
12	0	3	5	1	4	5	2	4	1	4.6	2.333333
13	0	2	5	1	4	5	2	6	1	2	4
14	0	3	5	1	4	5	2	5	1	5	4.666667
15	0	2	4	2	1	4	2	3	1	4.6	4.666667
16	1	2	1	2	1	5	2	5	1	3.2	2.666667
17	0	3	6	1	4	3	2	5	4	3.8	4.666667
18	0	2	5	1	4	5	2	6	1	1.8	2.333333
19	0	2	5	2	4	4	2	3	1	4.6	5
20	1	3	5	1	4	5	3	1	1	4.4	5
21	1	5	1	4	4	6	6	3	1	4.6	4.333333
22	1	3	1	3	4	5	1	2	1	4	3
23	0	4	5	3	4	5	2	4	1	3.2	4.333333
24	1	3	5	2	4	4	3	3	1	5	5
25	1	2	6	2	1	5	2	3	1	4	4
26	0	2	1	2	4	5	2	5	1	4.6	4.333333
27	1	3	1	2	4	5	2	6	1	4	4
28	1	3	1	1	4	4	2	3	1	4.2	4.333333
29	0	4	5	2	4	3	2	5	1	3.6	4
30	1	2	1	2	1	6	2	2	1	5	4.333333
31	1	3	1	1	4	5	7	1	1	3.6	4
32	1	6	5	4	4	5	6	3	1	4.6	4
33	0	3	1	1	4	5	3	2	1	4.8	4.666667

Fig. 10.1 Sample DCE data

To create a bar plot of a particular demographic variable (i.e., Marriage Status), we can execute the following R code:

```
demo_data <- read.csv("D:/R/DCE_Book_demo.csv")
barplot(table(demo_data$MarriageStatus),
    main = "Marriage Status Distribution",
    xlab = "Category",
    ylab = "Frequency",
    names.arg = c("Single","Married","Divorced","Widowed"))
```

Here is a breakdown of the code that we use:

1. **read.csv("D:/R/DCE_Book_demo.csv")**: This reads our dataset that is saved in a CSV file format
2. **table(demo_data$MarriageStatus)**: This creates a frequency table of the "MarriageStatus" variable in the "demo_data" data frame.

3. **barplot()**: This creates a barplot of the frequency table.
4. **main = "Marriage Status Distribution"**: This sets the main title of the plot to "Marriage Status Distribution".
5. **xlab = "Category"**: This sets the label of the x-axis to "Category".
6. **ylab = "Frequency"**: This sets the label of the y-axis to "Frequency".
7. **names.arg = c("Single","Married","Divorced","Widowed")**: This sets the names of the categories on the x-axis to "Single", "Married", "Divorced", and "Widowed".

The bar plot will be printed on the RStudio pane under "Plots" (as shown in Fig. 10.2):

We can also create a histogram for the variable "Level of environmental concerns" using the R code below:

```
demo_data <- read.csv("D:/R/DCE_Book_demo.csv")
hist(demo_data$Env,
   main = "Envrionmental Concerns Distribution",
   xlab = "Category",
   ylab = "Frequency")
```

The printed histogram is shown in Fig. 10.3:

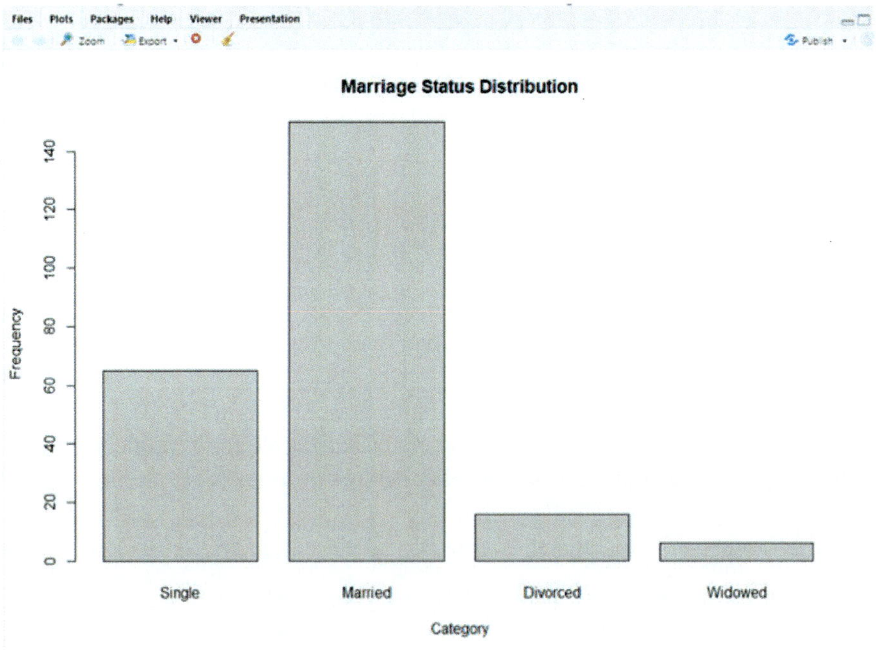

Fig. 10.2 A bar plot for the marriage status variable

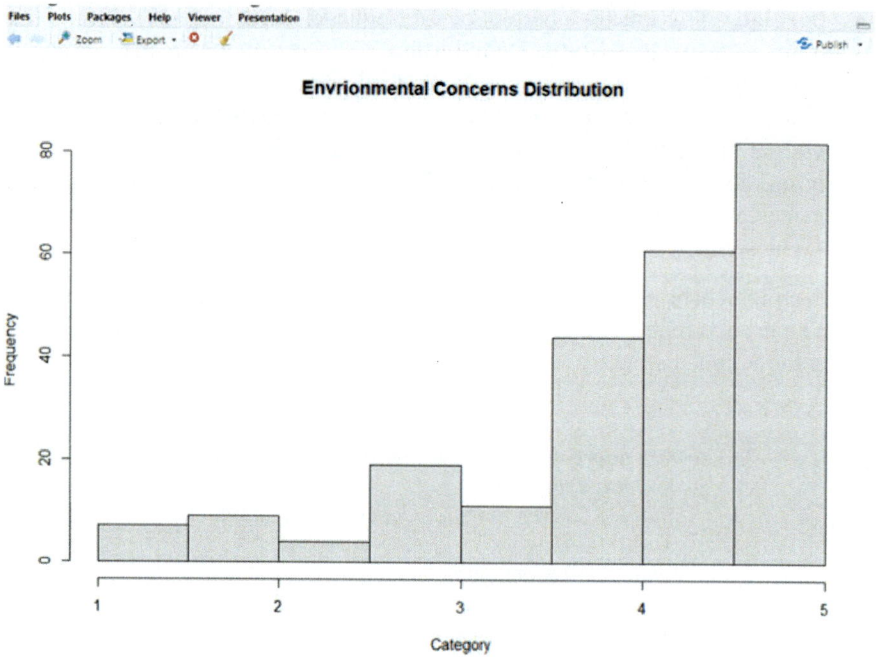

Fig. 10.3 A histogram for the environmental concerns variable

In addition, we can also visualize two variables at the same time using a box plot. For example, we visualized the variables Income Level and Age using the following R code:

```
demo_data <- read.csv("D:/R/DCE_Book_demo.csv")
boxplot(Age ~ IncomeLevel, data = demo_data,
    main = "Age Distribution by IncomeLevel",
    xlab = "Age",
    ylab = "Income Level")
```

The box plot will be printed on the RStudio pane as shown in Fig. 10.4:

We can also be a bit creative in doing the visualization, for example, by creating a heatmap of correlations between the demographic variables:

```
library(corrplot)
demo_data <- read.csv("D:/R/DCE_Book_demo.csv")
corrplot(cor(demo_data), type="lower", order="hclust",
    col=colorRampPalette(c("red", "white", "blue"))(100))
```

The R code above uses the **corrplot** package to create a correlation plot of the variables in our sample dataset. The ***cor()*** function calculates the correlation matrix

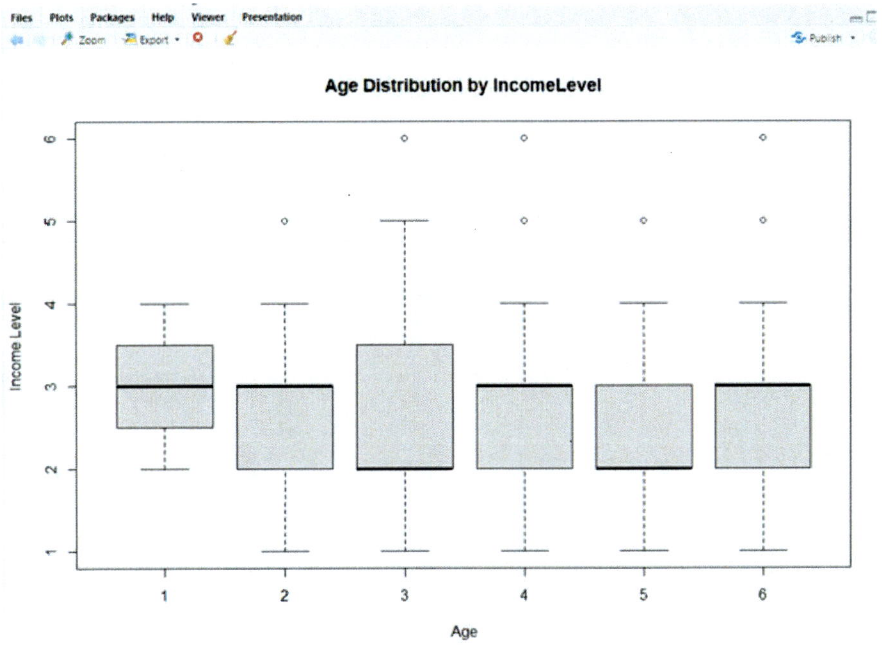

Fig. 10.4 A box plot for two variables

of the variables in the dataset, and the resulting matrix is passed to the ***corrplot()*** function to create the visualization. The heatmap will then be shown on the RStudio pane under "Plots" (see Fig. 10.5):

Alternatively, we can organize the correlation analysis into a standard correlation table in an APA style by using the package **apaTables**:

```
install.packages('apaTables')
library(apaTables)

demo_data <- read.csv("D:/R/DCEBook_Chapter10.csv")
cor_matrix <- cor(demo_data)
apa.cor.table(cor_matrix,filename="ex.CorTable1.doc")
```

A correlation table will be printed on the R console, as shown in the screenshot (Fig. 10.6):

The R code above also created a word file named ***ex.CorTable1.doc,*** saved in the working directory (see Fig. 10.7). The exported table can then be directly used in academic reports and papers.

Similarly, we can also create a table to summarize the demographic information of our DCE respondents using the R code below:

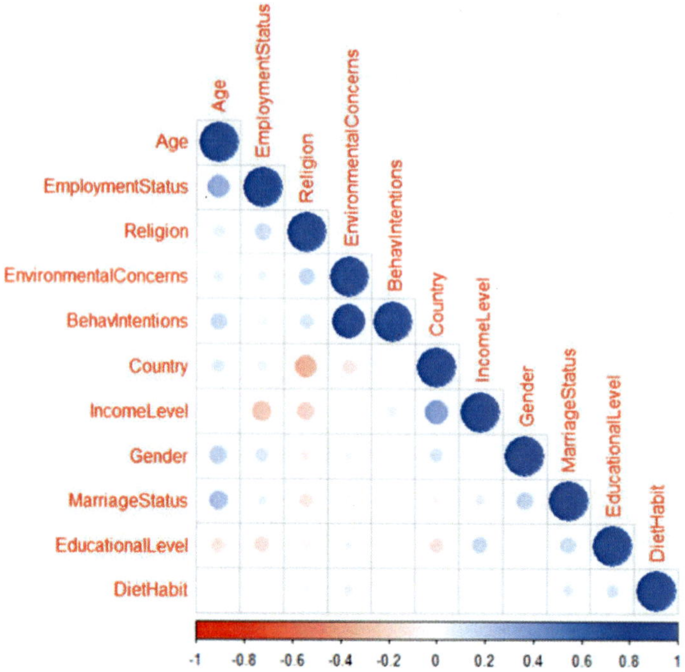

Fig. 10.5 A heatmap of correlation between demographic variables

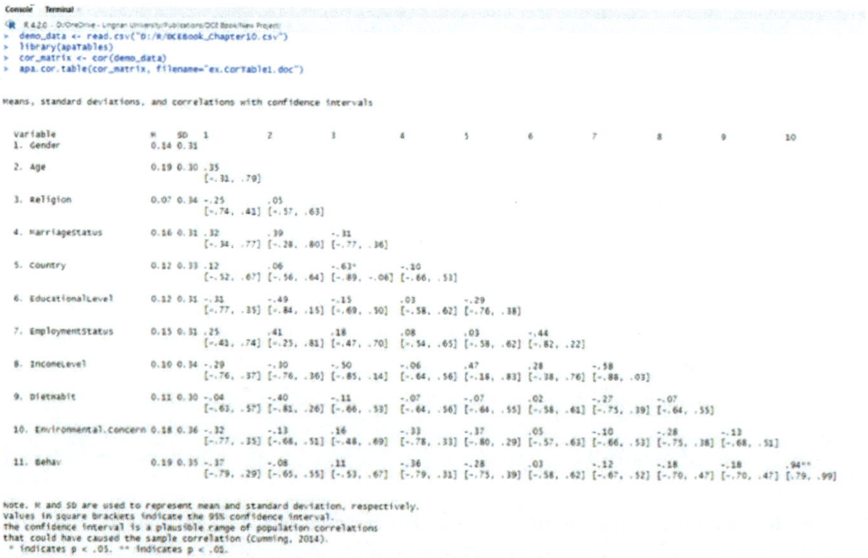

```
> demo_data <- read.csv("D:/R/DCEBook_Chapter10.csv")
> library(apaTables)
> cor_matrix <- cor(demo_data)
> apa.cor.table(cor_matrix, filename="ex.CorTable1.doc")
```

Means, standard deviations, and correlations with confidence intervals

Variable	M	SD	1	2	3	4	5	6	7	8	9	10
1. Gender	0.14	0.31										
2. Age	0.19	0.30	.35 [-.31, .70]									
3. Religion	0.07	0.34	-.25 [-.74, .41]	.03 [-.57, .63]								
4. MarriageStatus	0.16	0.31	.32 [-.34, .77]	.39 [-.28, .80]	-.31 [-.77, .36]							
5. Country	0.12	0.33	.12 [-.52, .67]	.06 [-.56, .64]	-.63* [-.89, -.06]	-.10 [-.66, .53]						
6. EducationalLevel	0.12	0.31	-.31 [-.77, .35]	-.49 [-.84, .15]	-.15 [-.69, .50]	.03 [-.58, .62]	-.29 [-.76, .38]					
7. EmploymentStatus	0.15	0.31	.25 [-.41, .74]	.41 [-.25, .81]	.18 [-.47, .70]	.08 [-.54, .65]	.03 [-.58, .62]	-.44 [-.82, .22]				
8. IncomeLevel	0.10	0.34	-.29 [-.76, .37]	-.30 [-.76, .36]	-.50 [-.85, .14]	-.06 [-.64, .56]	.47 [-.18, .83]	.28 [-.38, .76]	-.58 [-.88, .03]			
9. DietHabit	0.11	0.30	-.04 [-.63, .57]	-.40 [-.81, .26]	-.11 [-.66, .53]	-.07 [-.64, .56]	-.07 [-.64, .55]	.02 [-.58, .61]	-.27 [-.75, .39]	-.07 [-.64, .55]		
10. Environmental.concern	0.18	0.36	-.32 [-.77, .35]	-.13 [-.68, .51]	.16 [-.48, .69]	-.33 [-.78, .33]	-.37 [-.80, .29]	.05 [-.57, .63]	-.10 [-.66, .53]	-.28 [-.75, .38]	-.13 [-.68, .51]	
11. Behav	0.19	0.35	-.37 [-.79, .29]	-.08 [-.65, .55]	.11 [-.53, .67]	-.36 [-.79, .31]	-.28 [-.75, .39]	.03 [-.58, .62]	-.12 [-.67, .52]	-.18 [-.70, .47]	-.18 [-.70, .47]	.94** [.79, .99]

Note. M and SD are used to represent mean and standard deviation, respectively.
Values in square brackets indicate the 95% confidence interval.
The confidence interval is a plausible range of population correlations
that could have caused the sample correlation (Cumming, 2014).
* indicates p < .05. ** indicates p < .01.

Fig. 10.6 Printed correlation table

Table XX

Means, standard deviations, and correlations with confidence intervals

Variable	*M*	*SD*	1	2	3	4	5	6	7	8
1. Gender	0.14	0.31								
2. Age	0.19	0.30	.35 [-.31, .79]							
3. Religion	0.07	0.34	-.25 [-.74, .41]	.05 [-.57, .63]						
4. MarriageStatus	0.16	0.31	.32 [-.34, .77]	.39 [-.28, .80]	-.31 [-.77, .36]					
5. Country	0.12	0.33	.12 [-.52, .67]	.06 [-.56, .64]	-.63* [-.89, -.06]	-.10 [-.66, .53]				
6. EducationalLevel	0.12	0.31	-.31 [-.77, .35]	-.49 [-.84, .15]	-.15 [-.69, .50]	.03 [-.58, .62]	-.29 [-.76, .38]			
7. EmploymentStatus	0.15	0.31	.25 [-.41, .74]	.41 [-.25, .81]	.18 [-.47, .70]	.08 [-.54, .65]	.03 [-.58, .62]	-.44 [-.82, .22]		
8. IncomeLevel	0.10	0.34	-.29 [-.76, .37]	-.30 [-.76, .36]	-.50 [-.85, .14]	-.06 [-.64, .56]	.47 [-.18, .83]	.28 [-.38, .76]	-.58 [-.88, .03]	
9. DietHabit	0.11	0.30	-.04 [-.63, .57]	-.40 [-.81, .26]	-.11 [-.66, .53]	-.07 [-.64, .56]	-.07 [-.64, .55]	.02 [-.58, .61]	-.27 [-.75, .39]	-.07 [-.64
10. Environmental.Concern	0.18	0.36	-.32 [-.77, .35]	-.13 [-.68, .51]	.16 [-.48, .69]	-.33 [-.78, .33]	-.37 [-.80, .29]	.05 [-.57, .63]	-.10 [-.66, .53]	-.28 [-.75

Fig. 10.7 Exported correlation table

```
data <- read.csv("D:/R/DCE 1_demo2.csv")

library(gtsummary) #please note that these two packages should be installed first before they can be loaded
 library(dplyr)

 # Read data
 df <- read.csv("D:/R/DCE 1_demo2.csv")

 # Create summary table
 summary_table <- df %>%
  select(Gender, Age, Religion, MarriageStatus, Country, EducationalLevel, EmploymentStatus, IncomeLevel, DietHabit, Environment, Behav) %>%
  tbl_summary() %>%
  bold_labels() %>%
  italicize_levels()

 # Print summary table
 summary_table
```

After executing the code, a demographic table will be shown in the RStudio pane under the Viewer section. We can replace the numbers (e.g., ‘1’, ‘2’, ‘3’) with real values of each variable (see Fig. 10.8).

Fig. 10.8 Printed demographic table

Characteristic	N = 237[1]
Gender	109 (46%)
Age	
1	16 (6.8%)
2	93 (39%)
3	80 (34%)
4	24 (10%)
5	20 (8.4%)
6	4 (1.7%)
Religion	
1	134 (57%)
2	4 (1.7%)
3	4 (1.7%)
4	9 (3.8%)
5	55 (23%)
6	31 (13%)
MarriageStatus	
1	65 (27%)
2	150 (63%)
3	16 (6.8%)
4	6 (2.5%)

The second part of data visualization concerns the core DCE data and results. After we compute the preference coefficients, we can use the R package **stargazer** to convert the results into a nicely organized table format that can be quickly inserted into the report (Hlavac, 2018).

For instance, we demonstrate below how to use the R packages **support.CEs** and **survival** to analyze DCE data and estimate respondents' preferences. Once we have the results, we can execute the following R code to create the results table:

```
library(support.CEs)
desmat2 <- make.design.matrix(choice.experiment.design =
unlabeled, optout = TRUE,
                          categorical.attributes = c("Additives",
"Nutrition", "Convenience", "CarbonFootprint", "VeganLabel"),
                          continuous.attributes = c("Price"),
                          unlabeled = TRUE)
dataset2 <- make.dataset(respondent.dataset = df,
                      choice.indicators = c("q1", "q2", "q3", "q4", "q5",
"q6", "q7", "q8"),
                      design.matrix = desmat2)

library(survival)
clogout1 <- clogit(RES ~ ASC + Yes + Forty + Ready + Price +
Twenty + Yes.2 + strata(STR), data = dataset2)
clogout1

library(stargazer)
stargazer(clogout1, type = "html", out = "coefs.html")
```

Then, an HTML table of the model coefficients will be created and saved in a file called "coef.html" to the working directory of R (see Fig. 10.9 for a screenshot of the file).

We can also use the R package **ggplot2** to visualize the coefficients as a bar chart for a more straightforward visualization of DCE results. To do this, we can use the following R code after installing the package:

```
library(ggplot2)
clogit_model <- clogit(RES ~ Yes + Forty + Ready + Price +
Twenty + Yes.2 + strata(STR), data = dataset2)

# Extract the coefficients
coefs <- coef(clogit_ model)[1] #Remove ASC from the coefficients

# Plot the coefficients as a bar chart
ggplot(data.frame(attribute = names(coefs), coef = coefs), aes(x =
attribute, y = coef)) +
  geom_bar(stat = "identity") +
  xlab("Attribute") +
  ylab("Coefficient") +
  ggtitle("Coefficients of Conditional Logit Model")
```

In the R code above, the formula used for ***clogit()*** specifies the response variable (**RES**) and the predictor variables (**Yes**, **Forty**, **Ready**, **Price**, **Twenty**, and **Yes.2**). These predictor variables represent the attribute levels of the categorical attributes (**Additives**, **Nutrition**, **Convenience**, **CarbonFootprint**, and **VeganLabel**) and the

Fig. 10.9 A screenshot of the generated HTML file

	Dependent variable:
	RES
ASC	2.937***
	(0.176)
Yes	-0.116**
	(0.048)
Forty	0.139***
	(0.048)
Ready	-0.095**
	(0.048)
Price	-0.110***
	(0.027)
Twenty	0.145***
	(0.048)
Yes.2	-0.061
	(0.048)
Observations	5,688
R^2	0.167
Max. Possible R^2	0.519
Log Likelihood	-1,564.585
Wald Test	500.880*** (df = 7)
LR Test	1,036.768*** (df = 7)
Score (Logrank) Test	789.921*** (df = 7)
Note:	*p<0.1; **p<0.05; ***p<0.01

continuous attribute (**Price**) in the choice experiment. The first attribute levels were omitted in the dummy coding of the function ***clogit()***. Finally, the ***ggplot()*** function is used to plot the coefficients as a bar chart, with the attribute level names (**Yes**, **Forty**, **Ready**, **Price**, **Twenty**, and **Yes.2**) on the x-axis and the estimated coefficients on the y-axis.

The bar chart will be printed as shown in Fig. 10.10:

Moreover, we can also use the package **ggplot2** to plot the relative importance of each attribute by running the following R code:

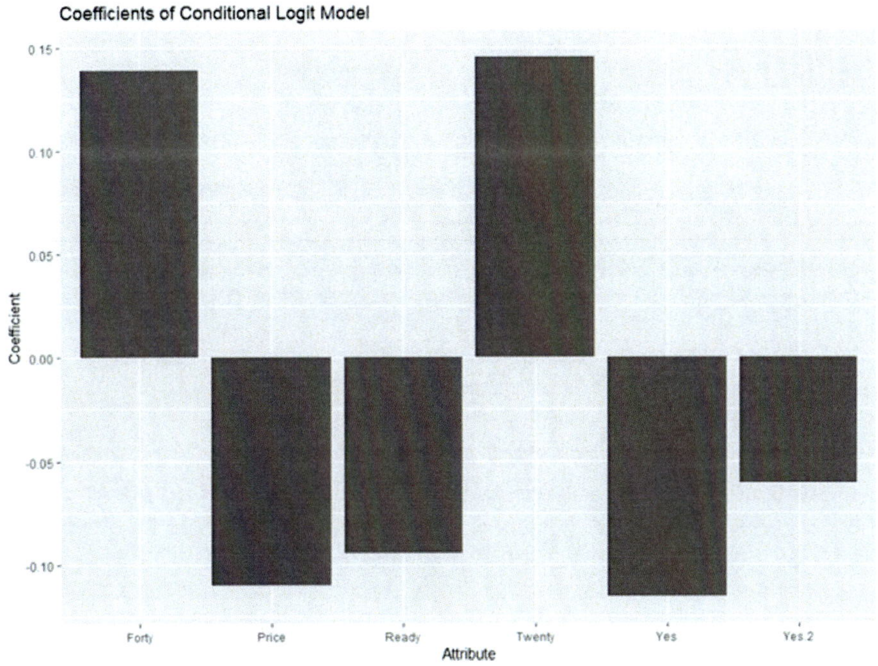

Fig. 10.10 A bar chart of DCE coefficients

```
library(ggplot2)

attribute_names <- c("Additives", "Nutrition", "Convenience",
"Price", "CarbonFootprint", "VeganLabel") #here, we replaced
the attribute level names with actual attribute names for easier
understanding

relative_importance <- exp(coefs)
df <- data.frame(attribute = attribute_names,
                 importance = relative_importance)
ggplot(df, aes(x = attribute, y = importance)) +
  geom_bar(stat = "identity") +
  ggtitle("Relative Importance of Attributes") +
  xlab("Attribute") +
  ylab("Relative Importance")
```

A bar chart that illustrates the relative importance of each attribute included in our DCE will then be created (see Fig. 10.11):

Additionally, we can also use the ***qqnorm()*** and ***qqline()*** functions in R to visualize the residuals of our conditional logit model. The visualization will show us how closely the residuals match a normal distribution, which is a key assumption of the conditional logit model (Boots & Kanaroglou, 1988). To calculate the residuals and create a Q–Q plot of the residuals, we can use the following R code:

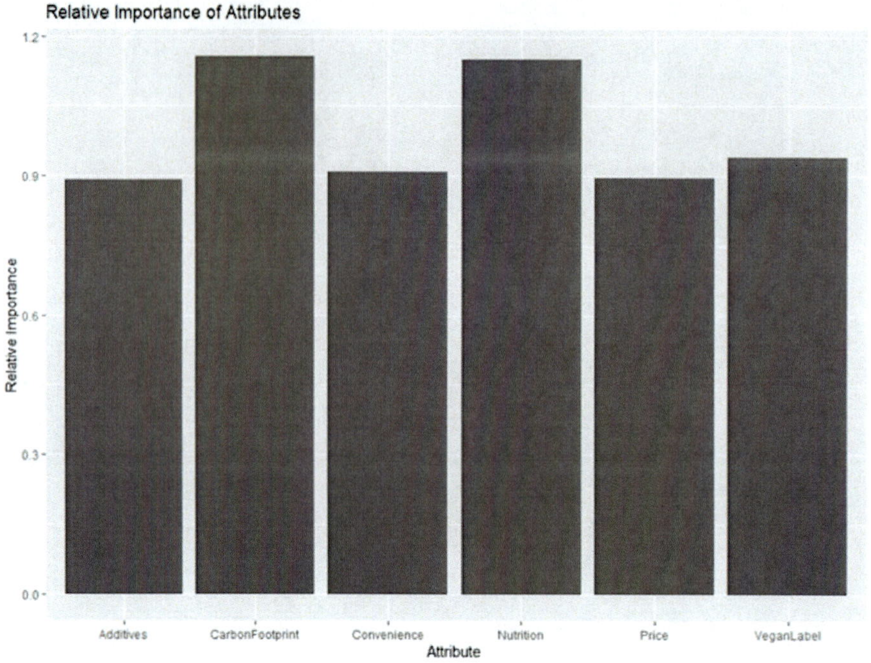

Fig. 10.11 A bar chart of the relative importance of attributes

```
# Calculate the residuals from your conditional logit model
res <- residuals(clogout1)
# Create a Q-Q plot of the residuals
qqnorm(res)
qqline(res)
```

After running the code, the Q–Q plot will then be printed on the RStudio pane as follows (see Fig. 10.12).

Based on the printed Q–Q plot, we can visually inspect to assess whether the residuals of the model follow a normal distribution. If the points on the plot deviate significantly from the straight line, it may indicate that the residuals are not normally distributed, which could affect the validity of the model's results. In our case, the plot of the residuals is not too far away from the line, indicating that the model's errors are random and unbiased and, thus the model is a good fit for the data.

10.2 Reporting the DCE Findings

To report the DCE findings, several key elements need to be included: (1) a description of the data and the sample (e.g., age, gender, employment status, educational level), (2) data analysis methods and steps (e.g., the selected model for estimation), (3)

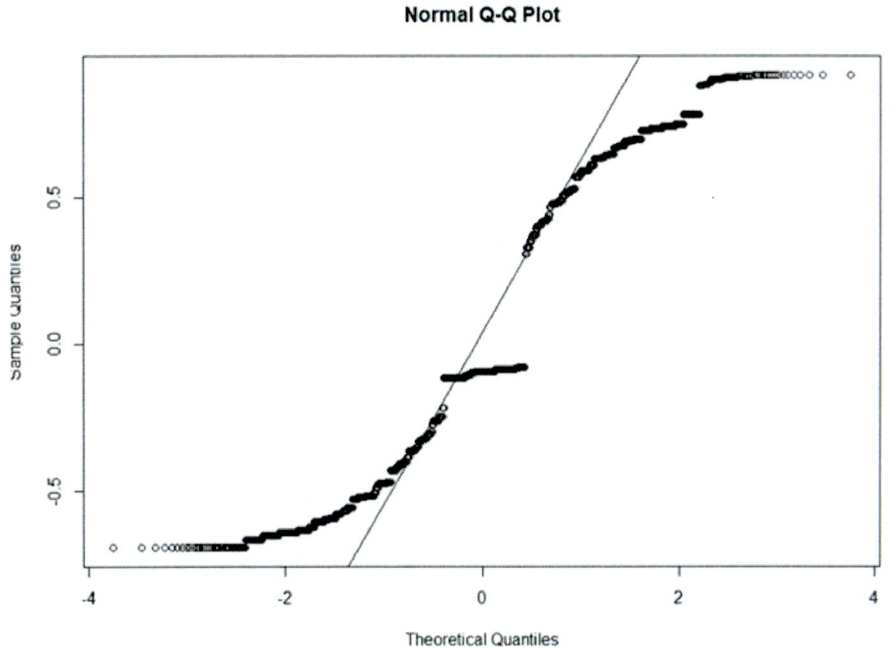

Fig. 10.12 A Q–Q plot of the residuals of the conditional logit model

the results of the model estimation (e.g., the coefficients and standard errors of the attributes, the model fit statistics, and the goodness of fit measures), and (4) attribute importance (e.g., the relative importance of attributes and the marginal WTP estimates). We will use our sample project on preferences for alternative meat products as an example to briefly illustrate how to report DCE findings in Textbox 10.1.

Textbox 10.1. An Example of DCE Findings Reporting

Results

We received a total of 255 responses with 237 valid responses (i.e., invalid responses include those incomplete and those that failed to pass the attention check question). Of the 237 valid participants, 128 (54.01%) were female and 93 (39.24%) of them were aged between 26 to 35. Slightly over half (56.54%) of respondents were Christians and around 78% of respondents had a Bachelor's degree or above. Across the sample, 150 (63.29%) of them were married and 79.32% were employed full-time. Moreover, almost 60% of our respondents had a monthly income level of over US$2,001. Table 10.1 summarizes the demographic information of the respondents in our DCE study.

Table 10.1 Demographic characteristics of respondents (DCE 1)

	Respondents ($n = 237$)	
Gender		
Male	109	45.99%
Female	128	54.01%
Age		
18–25	16	6.75%
26–35	93	39.24%
36–45	80	33.76%
46–55	24	10.13%
56–65	20	8.44%
66 or above	4	1.69%
Religion		
Christian	134	56.54%
Buddhist	4	1.69%
Jewish	4	1.69%
Muslim	9	3.80%
Nothing in particular	55	23.21%
Others	31	13.08%
Educational level		
Primary school or below	1	0.42%
Secondary school	6	2.53%
Diploma	20	8.44%
Associate degree/High diploma	23	9.70%
Bachelor's degree	146	61.60%
Postgraduate or above	41	17.30%
Marriage status		
Single	65	27.43%
Married	150	63.29%
Divorced	16	6.75%
Widowed	6	2.53%
Employment status		
Student	3	1.27%
Employed full-time	188	79.32%
Employed part-time	28	11.81%

(continued)

Table 10.1 (continued)

	Respondents ($n = 237$)	
Gender		
Temporary worker	1	0.42%
Seeking opportunities	6	2.53%
Retired	6	2.53%
Others	5	2.11%
Monthly income level (in USD)		
< 500	8	3.38%
500–1000	37	15.61%
1001–2000	63	26.58%
2001–3000	44	18.57%
3001–4000	40	16.88%
> 4000	45	18.99%

The McFadden conditional logit model was used in this study to analyze DCE data, as this model describes the attributes and the attribute levels of the alternatives and allows a small sample size comparable to that of other models (Hauber et al., 2016). The utility coefficients were calculated using the ***clogit*** function in an R package called **supported.CEs** (Aizaki, 2012). Parameter estimation was conducted for the total sample ($n = 237$) first, and the interaction effect of employment-related interventions with respondents' personal characteristics (i.e., age, gender) was also considered to understand how consumers' personal characteristics may affect their preferences for alternative meat products. Moreover, we calculated the goodness-of-fit of the models to evaluate the estimated CL models using McFadden's R^2 or pseudo R^2, which can be defined by the log-likelihoods of the model with and without predictors (Aizaki, 2012).

The results of mixed logit analysis for the main effects are provided in Table 10.2.

In general, the coefficient (coefficient here refers to the mean preference estimates) with the greatest magnitude was **Carbon Footprint**, followed by ***Nutrition*** and ***Additives***. The estimated mean for a 20% reduction in greenhouse gas emission ($\beta = 0.15$, $p < 0.001$) was statistically significant, suggesting that respondents were, on average, more likely to choose an environmentally friendly alternative meat product.

The remaining coefficients followed our expected trends, which demonstrated that respondents, in general, preferred alternative meat products with **higher nutritional values** ($\beta = 0.14, p < 0.001$), **without additives** ($\beta = 0.12, p < 0.05$), **with low-price** ($\beta = 0.11, p < 0.001$), and in a raw format rather than in a ready-to-eat format ($\beta = -0.09, p < 0.05$). However, we found that a vegan label was not a significant factor in the respondents' decision.

We calculated the relative monetary value of each attribute for ease of comparison (see the WTP values along with their confidence intervals in Table 10.2). These values indicate the relative importance of attributes of alternative meat products against each other. The WTP values showed that respondents were willing to pay an extra US$1.32 for an alternative meat product that reduces greenhouse gas emission, which is also top ranked WTP value among all attributes. High nutritional value and additives-free were also highly favoured among participants, who were willing to pay US$1.26 and US$1.05 for buying an alternative meat product that contains 40 g of protein and does not contain any additives respectively.

Table 10.2 Main Effects and WTP

Main effects-only and WTP				
Attributes	Levels	Coefficients	WTP ($)	95% CI
ASC		2.94**		
Addictives	Without addictives			
	With addictives	−0.12	−1.05	−0.34
Nutrition	25 g			
	40 g	0.14***	1.26	2.47
Format	Raw			
	Ready-to-eat	−0.09*	−0.86	−0.15
Price	US$3.5			
	US$4.5			
	US$6	−0.11***		
Carbon footprint	No info on reduction in greenhouse gas emission			
	20% reduction in greenhouse gas emission			
Vegan label	Without vegan label	0.15***	1.32	2.53
	With vegan label	−0.06	−0.55	0.15

Adjusted rho-squared: 0.25
Number of events: 1896
Log likelihood function: −2082.97
*Significant variables ($p < 0.05$)
***Significant variables ($p < 0.01$)
Log likelihood ratio test ($p < 0.00$)

We further conducted interaction analyses by including the personal characteristics of respondents (i.e., gender, age, educational level, monthly income level) and their levels of environmental concerns, pro-environmental behavioral intentions, and attention to the environmental impacts of food choices in the regression analyses. Our main findings for the interaction effects were summarized in Table 10.3. Due to a large number of interaction items included, we only presented statistically significant interations in the table. The analysis revealed that there is only one interaction effect that reached significance at the 95% confidence level.

Table 10.3 Main Effects with Interactions

With interactions		
Attributes	Levels	Coefficients
ASC		2.94***
Addictives	Without addictives	
	With addictives	−0.43***
Nutrition	25 g	
	40 g	0.14***
Format	Raw	
	Ready-to-eat	−0.10*
Price	US$3.5	
	US$4.5	
	US$6	−0.11***
Carbon footprint	No info on reduction in greenhouse gas emission	
	20% reduction in greenhouse gas emission	
Vegan label	Without vegan label	0.15***
	With vegan label	−0.06
Age	Additives: Yes: Age	0.11***

Adjusted rho-squared: 0.25
Number of events: 1896
Log likelihood function: −2082.97
*Significant variables ($p < 0.05$)
***Significant variables ($p < 0.01$)
Log likelihood ratio test ($p < 0.00$)

Our regression results suggested that only one personal characteristic had impact on individuals' decisions on alternative meat products, which is the age of respondents. In particular, the positive coefficient of the interaction between Additives and Age ($\beta = 0.11$, $p < 0.001$) indicated that the ***older the respondents*** were, the more utility they attached to alternative meat products that *contain* additives, compared to younger respondents.

References

Aizaki, H. (2012). Basic functions for supporting an implementation of choice experiments in R. *Journal of Statistical Software, 50*, 1–24.

Boots, B. N., & Kanaroglou, P. S. (1988). Incorporating the effects of spatial structure in discrete choice models of migration. *Journal of Regional Science, 28*(4), 495–510.

Hauber, A. B., González, J. M., Groothuis-Oudshoorn, C. G., Prior, T., Marshall, D. A., Cunningham, C., ... & Bridges, J. F. (2016). Statistical methods for the analysis of discrete choice experiments: a report of the ISPOR conjoint analysis good research practices task force. *Value in health, 19*(4), 300–315.

Hlavac, M. (2018). Stargazer: Well-formatted regression and summary statistics tables. *R Package Version, 5*(2), 2.

Sarkar, D. (2008). Multivariate data visualization with R. *Use R*.

Sievert, C. (2020). *Interactive Web-Based Data Visualization with R, plotly, and shiny*. Chapman and Hall/CRC. ISBN 9781138331457. Retrieved from https://plotly-r.com

Wickham, H. (2011). ggplot2. *Wiley Interdisciplinary Reviews: Computational Statistics, 3*(2), 180–185.

Chapter 11
Conclusion

This book covers the fundamentals of discrete choice experiments (DCEs), from experimental design to reporting DCE findings, and the various methods and techniques used in stated preference research of which DCE is a part. We have compared various stated preference methods and highlighted the strengths and benefits of using DCE in social and behavioral sciences research.

Our goal in this book is to demonstrate R as a powerful computing platform and how it can be used to support the implementation and analysis of DCE using a variety of R packages. R provides various tools for designing and conducting DCE experiments, from choice task generation and questionnaire design to estimation of choice models and visualization of results. Using step-by-step instructions from a sample project, we have shown how R could make it easier for researchers to design and conduct DCE studies.

The use of R in DCE research brings many benefits. First, R is very flexible and provides great versatility, which means that researchers can customize their experiments according to their specific needs and preferences. In particular, researchers can choose from a wide range of R packages (e.g., **support.CEs**, **idefix**, and **choiceDes**) that support different aspects of DCE, including choice task generation, estimation of choice models, and visualization of results. Moreover, using R can largely improve DCE design and implementation efficiency while reducing the risk of errors because of its scripting ability and automation capabilities. Another key advantage of using R in DCE research is that R can manage large datasets and perform complex calculations since it has a variety of statistical analysis packages and powerful data handling capabilities.

In addition, thanks to its open-source status and the vibrant R developer community who keeps updating and enhancing R statistical computing resources, R provides an accessible and affordable option for postgraduate students, researchers in academia and industry, and all other individuals interested in applying DCEs. This also means that researchers can easily share their code and results with others, enabling greater collaboration and sharing of information among researchers.

L. Shang and Y. Chandra, *Discrete Choice Experiments Using R*,
https://doi.org/10.1007/978-981-99-4562-7_11

In conclusion, this book provides a comprehensive overview of the various tools and techniques available in R for DCE research and is intended as a resource for researchers interested in using R for their stated preference studies. Using R in DCE research is an exciting and rapidly developing field with enormous potential for future growth and development. Whether you are a seasoned researcher or just starting, we hope this book will help you explore the full potential of R for your DCE studies.

Like all research methods and tools, DCE is not a panacea for all types of research. DCE is more appealing to certain types of research questions and traditions. For example, DCE research usually includes multiple variables (e.g., between 5 and 7 independent variables at multiple levels of attributes for each variable) in a single study; in some way, it resembles survey experiments. While this is considered a strength of DCE, mainstream experimentalist researchers still prefer parsimonious design using lesser variables (e.g., one independent variable, one dependent variable, and either a moderator or mediator or both moderator and mediator) to obtain "cleaner" results in examining people's choice behavior. Consequently, the DCE method is considered as not providing clean results. However, the argument against this is that DCE is a *more realistic* (and '*real*-time decision-making study') study that reflects the reality since decision-makers often need to make choices out of multiple options (e.g., choosing job A vs. B vs. C) that involve multiple attributes (e.g., low vs. medium vs. high salary, kind vs. tough supervisors, near vs. far locations, etc.).

Secondly, most published DCE research to date is conducted without using a theoretical model (using arrows and boxes to signify relationships and variables) and some even without using hypotheses. Hence, this creates a perception that DCE is more of an exploratory method for theory exploration rather than for theory testing. One reason for the limited use of theoretical models and hypotheses in DCE research could be the nature of the method itself. DCE is often used to explore consumer preferences and decision-making processes, which existing theoretical models may not fully understand or describe. To date, there is limited methodological development that implements DCE to test a research model or theory. Implicitly, the interaction effect is inherently studied and captured in DCE results via interaction analysis, which can be performed in R (see Sect. 9.3.1 in Chap. 9). Hence, the interaction hypothesis can be studied using DCE. However, testing process mechanisms in a theory or research model, particularly doing interaction analysis, is still rare in DCE studies. This is a fertile area for future research and methods development.

Thirdly, while DCE research has become increasingly popular in the social sciences, the use of a mixed-method design that incorporates DCEs and other research methods is not yet widespread. However, adopting such an approach can bring many benefits. Researchers can better understand the context and factors influencing participants' choices by combining DCEs with other methods, such as surveys or interviews. For instance, qualitative research methods can provide deeper insights into the underlying reasons behind participants' preferences and choices, while quantitative methods can provide a more precise estimation of the relative importance of different attributes. Additionally, using a mix of methods can increase the credibility and validity of the findings, as it allows for triangulation and cross-validation of results. Therefore, we encourage researchers to consider using mixed-method designs in future DCE studies.

Lastly, most published DCE research to date is built upon a one-shot study design. Given the importance of replication as a responsible scientific enterprise, using multiple DCE studies in a single study is urgently needed. In social sciences research that is influenced by psychological tradition, most studies use experiments in a building block approach where one experiment confirms the other experiments in a single project and multiple experiments are conducted, from a more simplistic (e.g., variables A and B) to more complex model (e.g., variables A, B, C and D) and from one empirical context (e.g., samples, countries) to other empirical contexts. Future research can easily tackle this issue by conducting multiple DCE studies in a single project to demonstrate the rigor of DCE and help establish the findings' validity and generalizability.